ON THE ORIGIN OF SEX

ON THE ORIGIN OF SEX

The Weird and Wonderful Science of How Our Planet is Populated

LIXING SUN

First published in Great Britain in 2026 by
Profile Books Ltd
29 Cloth Fair
London
EC1A 7JQ

www.profilebooks.com

Copyright © Lixing Sun, 2026

1 3 5 7 9 10 8 6 4 2

Printed and bound in Great Britain by
CPI Group (UK) Ltd, Croydon, CR0 4YY

The moral right of the author has been asserted.

All rights reserved. Without limiting the rights under copyright reserved above, no part of this publication may be reproduced, stored or introduced into a retrieval system, or transmitted, in any form or by any means (electronic, mechanical, photocopying, recording or otherwise), without the prior written permission of both the copyright owner and the publisher of this book.

A CIP catalogue record for this book is available from the British Library.

Our product safety representative in the EU is
BGC Sustainability & Compliance,
7 avenue du Général Leclerc, Paris, 75014, France.
https://baldwinglobalconsulting.com

ISBN 978 1 80522 328 3
eISBN 978 1 80522 330 6

For the three people whose contributions
to my life far exceed $r = 0.5$:

my parents, *Lijuan Zhu* and *Shangao Sun*,
and my sister, *Xiangyang (Sunny) Sun*.

Contents

Introduction 1

Part I: The Origin

1. A Ratchet, a Curse and a Story of Emergence – *The Origin of Sex* 9
2. Running with the Red Queen – *The Maintenance of Sex* 37
3. Male, Female and Beyond – *Sex and Diversity* 68

Part II: The Unfolding

4. The Makings and Variations of Sex – *Sex Determination and Development* 103
5. The Gender Bender – *Gender and Adaptation* 136

Part III: The Struggle

6. Red in Tooth and Claw – *Mate Competition* 171
7. A Taste for the Beautiful – *Mate Choice* 200
8. The Mating Game – *Sexual Conflict* 231
9. War and Peace in the Battle of the Sexes – *Sex and Evolutionary Destiny* 256

Epilogue 285

Acknowledgements 295
A Brief Guide to Key Terms 299
Notes 305
Image Permissions 344
Index 347

Note

In this book, scientific names are used only when a species has no widely accepted common name or when the common name is too vague to be specific.

INTRODUCTION

Tetrahymena thermophila isn't your typical single-celled organism if you're not familiar with protozoans. Under a microscope, it's a ciliated speedster, moving around like a pool cleaner on a caffeine high. No flashy name, no PR team – but trust me, it's unforgettable. You probably learned in school that cells stick to a simple plan: one nucleus, the central hub of life's instructions. But *T. thermophila** rewrites the rules. This little rebel has two nuclei, each with its own specialized job.

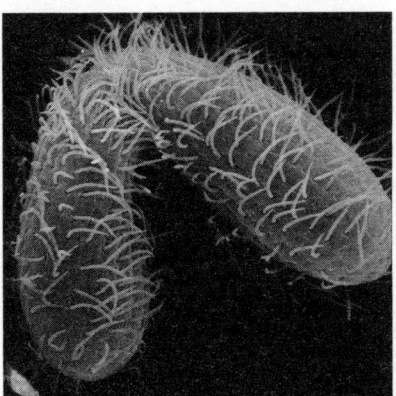

Figure 0.1. *Two Tetrahymena thermophila* caught mating

---

* As a rule in biology, *Tetrahymena* is abbreviated as *T.* after its first mention.

The larger nucleus is the steady workhorse, taking care of the basics: nutrition, growth and other essentials that keep the cell ticking. Meanwhile, the smaller nucleus has a more specialized gig: it's in charge of sex. Like most protozoans, *T. thermophila* usually have the simple life – copying itself through asexual cloning. But things take a turn when it decides it's time for a little romance (well, protozoan-style). That's when two cells snuggle up and get ready to swap their small nuclei. There's just one hard rule: no hooking up with someone of the same type. They can only pair with a mate sporting a different version of a particular gene. And this isn't a simple yes-or-no quiz; this gene comes in *seven* distinct versions, sparking the intriguing question: how many sexes does *T. thermophila* actually have?

The answer? One, because there's no 'male' or 'female' in sight. Their reproductive cells are the same in size and appearance. You can't tell who's bringing the sperm or eggs to the table. So they are unisexual. And yet they still go all in on sexual reproduction. Each of their seven gene variants acts like a 'mating type' – basically a kind of proto-sex. With seven of these in the mix, you get twenty-one possible pairings. That's one wildly complicated dating pool for a single-celled organism!

That means *T. thermophila* has a radically different mating system that makes our binary, mammalian notions of male and female look, frankly, boring. It's a jaw-dropping reminder that in nature, sex isn't just more diverse than we think – it's downright dazzling.

If a single-celled organism can pull off such a unique reproductive strategy, imagine what other surprises the natural world has in store. If we scale things up, we meet species like

Introduction

the European mole, a subterranean predator found across Asia and Europe. Each day, this voracious digger consumes a feast of earthworms and insects equal to half its body weight, but its diet isn't the most remarkable thing about it.

The female mole is a biological rule-breaker, a true rebel of the animal kingdom. She has both ovaries *and* testes, bundled into a remarkable package called ovotestes. That's right – she's packing the internal equipment we typically use to sort the biological sexes. Come spring, she takes on the role of mom with gusto, mating and giving birth to a wriggling litter of four or five pups. But after her maternal duties kick off, her ovaries take a backseat, shrinking down like fresh grapes drying into raisins, while her testes surge, running on a steroid high. The result is a testosterone explosion that transforms her into a territorial powerhouse, fiercely aggressive and ready to defend her pups against anything that dares to cross her burrow's threshold.

So what do we make of her? Is she truly female, male, or something in between? A hermaphrodite, perhaps?*

Through the course of this book we're going to tour some of the strangest, most enthralling wonders the natural world presents. We'll dive deep into the intricate, often mind-bending biological world of sex and gender in the animal kingdom and beyond. By exploring layers of hidden complexity, unravelling evolutionary mysteries and challenging common misconceptions, I hope to shed new light on these endlessly fascinating – and frequently misunderstood – scientific subjects.

---

* The answer is that although it has both ovarian and testicular tissues in ovotestes, it's still a female because it only produces eggs, not sperm. We'll meet the European mole again with more information later.

As you're about to discover, sex, from a biological perspective, unfolds across at least three dimensions: its origin, its development and its evolutionary path. In line with this, we will tour the world of sex through three parts, each devoted to one of these core themes: the emergence of sex, the mechanisms behind sex determination and development, and the powerful role of sexual selection in evolution.

Part I will take you back 2 billion years to explore the origin of sex, tracing how it became life's preferred mode of reproduction. Here we'll tackle the big questions: what's the point of sex? And why did it outshine cloning as the go-to strategy for complex organisms like plants and animals? Along the way, *On the Origin of Sex* unpacks what it actually means to be 'male' or 'female', explores the dazzling variety of sexes in nature, and examines why a two-sex system became so widespread in the first place. Most importantly, it shows how sex set the stage for what I call 'two forms of existence' (elaborated in Chapter 3) and genetic individuality – the spark behind life's incredible diversity.

Building on this foundation, Part II delves into the intricate biology of sex determination, exploring how different species create males and females and showing why intersex is an adaptive strategy. You'll meet species that can switch sexes to gain evolutionary advantages, turning the idea of fixed sex on its head, as well as animals that challenge human assumptions by bending gender roles to suit their needs.

Part III tackles sexual selection, where members of the same sex battle for mates while those of the opposite sex call the shots. *On the Origin of Sex* reveals how mating strategies adapt to social and ecological pressures, reshaping old tropes about 'coy ladies' and showcasing fierce female competition for mates. It uncovers a wealth of evidence showing that

Introduction

both males and females have mastered their own tricks of the trade to evolutionary success, breaking down biases and debunking any pseudoscientific basis for sexual stereotypes.

Now, let's embark on a journey into the world of sex – a world that stretches far beyond the boundaries of our *Umwelt*, a German word for the little world that our human senses can directly reach. Get ready for an experience that may feel both familiar and strange, even downright outlandish … not unlike sex itself.

Part I
THE ORIGIN

1

A RATCHET, A CURSE AND A STORY OF EMERGENCE

The Origin of Sex

The California condor, North America's largest bird, is a majestic vulture with a wingspan of over 10 feet, a body mass of up to 26 pounds and a lifespan of nearly sixty years. Revered as a sacred and mighty Thunderbird by Indigenous people such as the Yurok tribe, it rules the skies with power and awe. However, once a symbol of unbridled freedom, the condor now represents a poignant struggle for survival. Under the shadow of pollution, the condor's grandeur has dimmed as lead and DDT silently seep into its world, poisoning not only the bird but the spirit of wildness it embodies.

In 1987 the California condor hit rock bottom – it was declared extinct in the wild. With just twenty-seven birds left, conservationists took a bold gamble, bringing them into captivity as a last-ditch effort to save the species. Against all odds, their hard work paid off, and by 1991 these avian giants were returned to the skies, reclaiming their rightful place over California.

San Diego Zoo has been a key player in the effort to save the condor; for over three decades, scientists and staff members have worked like an avian matchmaker, pairing up compatible

On the Origin of Sex

Figure 1.1. The California condor

male and female birds and running DNA tests on blood, tissue, and eggshells to keep track of their genetic diversity.

This sounds like fairly routine bird business, but in 2021 the California condor made a new kind of stir – in the media, when two male condor chicks hatched from unfertilized eggs laid by two females.[1] For many, this discovery was like opening your refrigerator and finding chicks strolling out of an egg carton. It cracked open a long-guarded secret of the condor: females possess the miraculous ability of virgin birth, a remarkable testament to nature's persistent ability to surprise.

Virgin birth isn't just the stuff of ancient myths – it's a bizarre reality in the animal kingdom. Ever since domestic pigeons were caught doing it in 1924, rumours have been flying about female animals pulling off the same trick, popping out babies with no males in sight. For years, people brushed these stories off like sightings of the Loch Ness monster – until modern genetic technologies swooped in to prove sceptics wrong. Today, scientists have confirmed instances of virgin birth in over eighty vertebrate species,

A Ratchet, a Curse and a Story of Emergence

from bonnethead sharks and boa constrictors to whiptail lizards and Komodo dragons – and yes, the California condor.*

In science, virgin birth goes by names like cloning or parthenogenesis. While we'll be using these terms loosely in this book, it's worth knowing they're often not quite the same thing in biology. Both, however, belong to what biologists call asexual reproduction – basically, making babies solo.

So why do some creatures stick to the route of asexual reproduction while others dive into the wild, unpredictable world of sex? Or better yet – let's go straight to the heart of this chapter: where did sex come from in the first place?

At first glance, sex seems simple enough. But in biology it's anything but. In fact, the question 'Why sex?' is often called the queen of evolutionary problems – and not just because it sounds dramatic. It earns that royal title because it's one of the most puzzling, persistent mysteries in all of biology. What makes this question so special? Because, as you'll soon see, chasing its answer is both maddeningly difficult and irresistibly exciting.

. . .

To understand why sex evolved, we first need to imagine life without it. Asexual reproduction is everywhere – especially in the microbial world, where bacteria, archaea, protists and fungi rule. It's also common among invertebrates, algae and plants like liverworts and mosses. Asexual reproduction is the most straightforward way to make more life – no need for partners, courtship or mixing genes. But don't let its simplicity fool you. Even complex organisms, like flowering plants and some vertebrates, sometimes reproduce this way too.

---

* Mammals are incapable of parthenogenesis because embryo development requires essential genes from the sperm.

On the Origin of Sex

If you take a look at a vast, awe-inspiring bamboo grove or an aspen forest ... Surprise! Every single plant could be a clone of one ancestor that spread its roots far and wide. And it might sound a bit wild, but this cloning spree isn't too different from how a zygote divides into a deeply entwined and well-organized network of cells to create, for example, the human body. Sure, we don't usually count cell division – technically known as mitosis – as reproduction, but the cloning process is basically the same, and it's everywhere in nature.

With sexual reproduction appearing much later down the evolutionary pipeline than asexual reproduction, it can be tempting to see it as something of an upgrade. And yet asexual reproduction stuck around, even after sexual reproduction evolved across species. The reason for this is simple: it's efficient. Imagine evolution as a high-stakes race where every organism is vying for the ultimate trophy, that is, passing on as many of its genes as possible. When it comes to speed, nature hasn't invented a shortcut faster than asexual reproduction. It's the genetic fast lane. One solo organism becomes two, then four, then eight, and so on. Each generation doubles in number, creating an exponential growth pattern that looks like a hockey stick on a graph. This explosive process would keep going, too, if it weren't for pesky things like limited habitat and food shortages.

The advantage of asexual reproduction is obvious: every individual can create a new generation all on its own. No partner needed; no strings attached. In contrast, sexual reproduction demands two parents to achieve the same outcome. This means asexuals can, all things being equal, multiply at least twice as fast as their sexual counterparts. (Or, flipping the perspective, sexual reproduction is twice –

twofold, as preferred in biology – as costly as asexual reproduction.) The efficiency of asexual reproduction lies in skipping the step of producing males. If sex weren't a factor, all the resources spent on making males could go straight into producing more offspring, doubling the rate of population growth. This raises a cheeky question for biology: are males useless? When it comes to asexual reproduction, the answer is a definitive yes.

However, if asexual reproduction is the speedy shortcut, why do males still play such a critical role in the reproductive process? This puzzle, affectionately known as the 'cost of males', leaves scientists baffled. As evolutionary biologist John Maynard Smith famously put it, it's the 'twofold cost of sex' – a real eyebrow-raiser when you consider the material and energetic price tag attached to keeping males in the mix.[2]

When it comes to genetic efficiency, the 'cost of sex' and the 'cost of males' are not quite the same thing.* The twist lies in meiosis – a process where diploid body cells, each packed with two full chromosome sets (denoted $2n$), split, leaving sperm and egg cells with just one set (haploid, or n) before fertilization. So each parent only gets to pass down half of their genetic fortune to their offspring. Biologist George Williams dubbed this the 'cost of meiosis' back in 1971.[3] And as we dig into this genetic 'half-off' sale in Chapters 3 and 4, you'll see how it plays into the dizzying variety of sexual strategies – especially under the evolutionary force known as kin selection.

---

* Some biologists argue that the twofold cost of sex pertains specifically to the production of males, not cost of meiosis. (See Lehtonen, J., Jennions, M.D. and Kokko, H., 2012. The many costs of sex. *Trends in Ecology & Evolution*, 27(3), pp.172–8.)

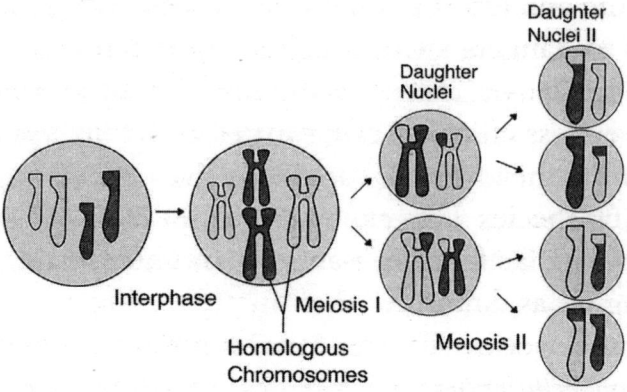

Figure. 1.2. The process of meiosis starts with one cell with two pairs of chromosomes ($2n$, with $n=2$) and ends with four cells, each with two chromosomes (n). Note that after Meiosis II, some gamete cells contain segments from both of the original chromosomes due to crossing-over or genetic recombination

Let's use humans to illustrate why sex is so inefficient. If you were a woman capable of asexual reproduction,* like the California condor, you wouldn't need to bother with producing eggs that contain only 23 chromosomes, half your genetic material. Instead, you could pass on all 23 *pairs* – 46 chromosomes – which are your entire genetic arsenal. But in sexual reproduction, meiosis only allows half of your genes (whether male or female) to pass on to the next generation. This slashes the efficiency of asexual reproduction in half.

In addition to this twofold cost of sex, sexual reproduction brings other challenges that can further reduce its efficiency. Finding, courting and mating with sexual partners

---

* While there is no such thing as male or female in asexually reproducing species, biologists often refer to asexuals as 'females' for convenience in discussion. The reasons for this will become more obvious later when we talk about parthenogenesis.

requires time and effort. It's no easy process, often exposing organisms to dangers such as fights, predation and sexually transmitted diseases. As a result, sexual reproduction can become even less efficient, driving up the already steep costs of sex even further.

In many species like cannibalistic spiders and praying mantises, sex is far from romantic. Males often face a grim fate, ending up as dinner for their mates during courtship or mating. Some species, like the Asian communal orb-weaving spider *Philoponella prominens*, take evasive action to avoid this very fate. Once mating is done, males catapult themselves away from females to avoid becoming a post-coital snack. Their escape is so speedy that it remained a mystery until an ultra-fast camera finally caught this process in action in 2022.[4]

Beyond the dramatic femme fatale encounters, battling for mates is a common and often brutal affair in nature. Watching elk, blackbirds, rattlesnakes, and even water striders during their mating seasons reveals just how tough it can get. Many of these animals end up exhausted, injured, mutilated, or worse – being killed by competitors or falling prey to predators in their quest to reproduce.

Consider the plight of salmon. After several years surviving in the open sea, they embark on a long and arduous journey back to their birthplace to spawn. It's a gruelling struggle and a tragic tale in many ways. Countless salmon become meals for predators like bears and eagles along the way, even before they make it to their breeding stream, much less have a chance to mate. Such is the high cost of sexual reproduction.

When we humans think about our own species, we can all agree that sexual relationships are among the most complex parts of our social lives. They can trigger every emotion imaginable: joy, inspiration, passion, commitment, devotion, self-sacrifice, obsession, sorrow, jealousy, betrayal, hatred. After all, half the world's songbooks are little more than case studies in love gone right – or spectacularly wrong.

Emotional swings aside, we all know there is a material cost of love as well. Yet despite long courtships, pricey diamonds and lavish weddings, more than half of marriages in many industrialized nations end in divorce. It makes you wonder: wouldn't it be easier to simply bypass intimacy by scraping off a few cheek cells to create our offspring?

Given the apparent challenges of sex, you might expect asexuals to have a clear evolutionary advantage. However, when we look at the natural world, especially in multicellular organisms like plants and animals, the opposite seems to be true. In fact, over 99.9 per cent of eukaryotes – organisms whose cells have the nucleus and other structures neatly wrapped in membranes – engage in sex, at least occasionally. So, despite being less efficient in reproduction, sexuals have not only persisted but have also thrived, becoming the norm rather than an exception. What unique qualities does sex possess that have led to its emergence and success?

Geneticist Hermann Joseph Muller probed into this 'why sex?' issue by flipping the question round: what does asexual reproduction, despite its maximal efficiency, do wrong? And in searching for the answer, he stumbled on a discovery that would later bear his name: Muller's ratchet.

• • •

Muller's claim to fame came with his groundbreaking discovery that X-rays could mess with our genes – a revelation that earned him a Nobel Prize. But even without that shiny medal, his legacy was sealed by his theory on the origin of sex. His core insight sprang from a lifelong obsession with mutations in the humble fruit fly, *Drosophila*. Driven by a relentless passion for scientific discovery, Muller's life became a whirlwind of twists and surprises, taking him from New York and Texas to Germany and Russia, and back again – each stop packed with wild adventures.

Born in 1890, Muller was fascinated by genetics during childhood. His path towards a scientific career took a significant leap when he earned a Cooper-Hewitt scholarship to attend Columbia University. After completing his Bachelor of Arts in Zoology, he set his sights on pursuing a Ph.D. under the eminent geneticist Thomas Hunt Morgan, celebrated for his groundbreaking studies of fruit fly chromosomes. However, financial limitations forced Muller to accept scholarships from Columbia and Cornell Medical College,* where he shifted gears to study physiology.†

Only two years later did Muller secure an assistantship from Columbia's Zoology Department, allowing him to join Morgan's famed Fly Room – Room 613 in Schermerhorn Hall – for his Ph.D. research. There, he explored how chromosomes swap genetic material through a process he named crossing-over (or genetic recombination), a critical feature of sexual reproduction whose significance Muller had already recognized at the time.

Surprisingly, Muller's earlier detour into medical school turned out to be a blessing in disguise. While studying

---

* Muller's father died of a cerebral stroke when he was just ten years old, leaving him to shoulder adult responsibilities early on. To support his mother and sister, he took on part-time jobs (especially as a bank clerk) during college and kept hustling with odd jobs throughout his graduate studies.
† Unless otherwise noted, all factual information in this section is based on Elof Axel Carlson's 1981 biography of Muller, *Genes, Radiation, and Society* (Ithaca, NY: Cornell University Press).

physiology, he encountered a problem that would ignite his lifelong obsession with mutations. After Wilhelm Röntgen's discovery of X-rays in 1895, these invisible rays quickly became a sensation in medical diagnostics. Doctors, nurses and lab technicians often used them with little protection. But Muller wasn't so easily swept up by the excitement. Instead, he questioned whether these seemingly miraculous rays could potentially trigger genetic mutations.

In 1915, after being recruited by the Rice Institute (now Rice University) in Houston, Texas, Muller set out to test his theory. Over the next three years, he exposed *Drosophila* to various doses of X-rays, meticulously documenting the types and numbers of mutants that appeared. Despite facing a series of job changes between 1918 and 1921, Muller never ceased his research on mutations. His relentless dedication led to several major discoveries, most notably establishing the connection between radiation and genetic mutation and confirming the dangers X-rays posed to living organisms.

Beyond his scientific work, Muller, like many intellectuals of his time, grew more and more critical of laissez-faire capitalism while teaching at the University of Texas at Austin. He even became an advisor to the communist National Student League on campus and took on the role of editor for *The Spark*, an underground magazine.

However, Texas, a conservative stronghold hostile to communist ideals, was a difficult environment for his political leanings. His leftist ideology left him increasingly isolated, and his personal life began to unravel. His wife, Jessie, a maths professor, was fired after the birth of their son, David, with her colleagues arguing that a mother couldn't be both a good teacher and a good parent. To make matters worse,

Muller's involvement in communist activities drew the attention of the FBI, who opened an investigation into him, adding yet another layer to his mounting troubles.

Under the weight of these and other pressures, he descended into paranoia, suspecting colleagues of stealing his ideas. Feeling overwhelmed and helpless, he tried to kill himself with barbiturates in an oak woodland outside Austin. But the suicide attempt failed. When discovered by a search team, he emerged from the wilderness 'walking, muddied, and wrinkled by an overnight rain, and somewhat confused'.[5]

Seeking refuge from personal and political turmoil, Muller left Texas – as well as his son and estranged wife – in 1932 for Berlin, thanks to a Guggenheim Fellowship. However his timing proved unfortunate, as Adolf Hitler's rise to power as chancellor in 1933 brought about the persecution of communists by the Nazis. Fortunately, an invitation from Nikolai Vavilov, Director of the Institute of Genetics, offered Muller an escape route to continue his research in Leningrad.*

Unaware of what he had got himself into, Muller immersed himself in his mutation studies upon arriving in the Soviet Union. But just as his research began to yield promising results, the horrors of the Great Purge of 1936–8 unfolded.† Stalin's brutal political repression swept across the nation, leading to the imprisonment, execution and exile of millions.

---

* The institute was moved from Leningrad (now St Petersburg) to Moscow in 1934.
† The Great Purge was Stalin's ruthless campaign of mass repression, designed to tighten his iron grip on the Soviet leadership and consolidate personal power.

Within this turmoil, the Institute of Genetics became a disaster zone under the oppressive rule of Stalin's acolyte in academia, Trofim Lysenko. Lysenko, a poorly educated quack masquerading as a geneticist, didn't even believe genes existed. But he wielded a tremendous amount of influence over scientific discourse throughout the Eastern bloc, extending as far as Mongolia and China. Tragically, among the casualties of the Great Purge were Vavilov and two brilliant geneticists, Solomon Levit and Isador Agol, whom Muller had brought to and mentored in Austin, Texas just a few years prior.

Watching the political system crumble into dictatorship and repression, Muller had a harsh wake-up call about the fatal flaw in the Soviet utopia. Seeing his colleagues and students arrested, jailed or simply disappeared, he realized he had to get out to save himself. In 1937 he took a sabbatical leave from the Institute of Genetics. He had no idea how close a call it was – Stalin, annoyed by Muller's book *Out of the Night*, was just about to target him.

How to escape, though? With limited options, Muller chose to volunteer for the Soviet-backed Republicans in the Spanish Civil War, fighting against the Nazi-supported Nationalists. In Madrid he served as medical staff alongside Norman Bethune, a Canadian doctor who would later die from a blood infection during surgery in China in 1939. However, before long the war began to turn against the Republicans and Muller found himself once again on the run.

It wasn't until 1945 that Muller finally found a permanent home at Indiana University in Bloomington. At fifty-five, the restless wanderer settled down at last. Remarkably, his passion for mutation research never wavered, even during his hardest times. Despite divorcing his wife and leaving his son

A Ratchet, a Curse and a Story of Emergence

behind in Texas, he always kept his most prized possession – his strains of fruit flies – close at hand. He even carried them while fleeing from the Soviet Union to the battlefields of Spain and from Europe back to America.

Muller's breakthrough discoveries about mutations – especially the alarming effects of X-rays on genes – became the defining feature of his career. Yet it wasn't until a year after the world had reeled from the atomic bombings of Hiroshima and Nagasaki in 1945 that his pioneering work was finally awarded a Nobel Prize.[6]

Since mutations are at the heart of evolution, it's no surprise that Muller – the man whose life was virtually synonymous with mutation – was among the first to realize that harmful mutations could accumulate in asexual lineages, which lack the ability to reverse this genetic decay. This process is much like what happens when a computer does not have regular system maintenance: errors, viruses and malware gradually accrue until the system's performance collapses.

Creatures that rely only on cloning pile up harmful mutations over time, a burden Muller called the 'mutation load'. Just like it sounds, this load drags down an organism's vitality, and if it keeps building generation after generation, it can drive the entire lineage to extinction – a genetic death sentence. This is the fatal flaw of asexual reproduction, no matter how efficient it seems. It helps explain why less efficient but more flexible strategies, like sex, have emerged and thrived.

In a 1964 article, just three years before his death, Muller coined the term 'ratchet' to describe this slow, degenerative genetic process as 'a serious long-run retardation of evolution' linked to asexual reproduction.[7] Geneticist Joe Felsenstein later honoured Muller's work by renaming the idea 'Muller's ratchet'. Today, it is a fundamental concept for any student interested in the origin of sex.*

· · ·

At this point you can see that asexuals, despite their high reproductive efficiency, are vulnerable to extinction due to Muller's ratchet. Over time they pick up harmful mutations, which increase their genetic burden. This genetic load acts like a ticking time bomb, leading eventually to a point known as 'mutational meltdown',[8] beyond which the entire asexual lineage crumbles into extinction. So how can these creatures escape the curse of Muller's ratchet?

Nature has several tricks up its sleeve. One avenue is through good mutations, which can promote your fitness. But there's a catch: random mutations are much more likely to be harmful than helpful. Counting on favourable mutations to reverse the ratchet is like expecting a good fox to guard your chicken coop instead of eating the chicken. So waiting for salvation from good mutations is unlikely to be viable.

Having a small genome is another way to dodge the curse. Within each domain of life – Bacteria, Archaea and Eukarya – the mutation rate per unit of genetic material is roughly the same. So the bigger the genome, the higher the chance of getting at least one harmful mutation, much like how you're more likely to make a typo the longer the essay you write. Conversely, organisms with smaller genomes have

---

* The role of Muller's ratchet in the origin of sex remains a subject of debate. Some evolutionary biologists propose that genes involved in meiosis may have been co-opted from genes for DNA repair. Thus repairing damaged DNA, rather than eliminating mildly harmful mutations, may offer a more compelling explanation for the origin of sex. (See elaborations later in the chapter.)

A Ratchet, a Curse and a Story of Emergence

a lower probability of getting a bad mutation. This might explain why mitochondria, with their tiny circular DNA of about 16,000 base pairs in most animals, can keep cloning themselves indefinitely.*

But organisms with tiny genomes face their own set of challenges, too. A genome that's too small can't carry all the essential genes required to live. Take the human mitochondrial genome, for example – it only has thirty-seven genes and can't survive on its own. So what's the minimum number of genes needed for independent life? Currently, the record-holder is an artificially engineered bacterium with a genome of just 473 genes.[9]

Small and shrinking genomes may also experience another setback: their constituent genes tend to desert them and move elsewhere. Genes are like snooty socialites, really. If they sense that things aren't looking good, they jump ship to find greener pastures.† For this, the mitochondrial genome is again a prime example of how genes pull off their great escape.

First off, the mitochondrion is a hostile place for genes because it's full of corrosive chemicals, leading to what biologists call oxidative stress.[10] The DNA there takes a beating by more mutations. So, in a desperate bid for survival, many genes in the mitochondrion made a mad dash to the safer neighbourhood – the cell's nucleus – throughout evolutionary history.[11] With about 96.5 per cent of the genes having already made their escape (compared to the gene count in

---

* Chloroplast DNA in green plants is similar, but its genome size is about 6.0–11.5 times larger than mitochondrial DNA. Also, selection against deleterious mutations in mitochondrial DNA keeps mitochondria viable.
† There are many ways to move genes around, including (but not limited to) genetic recombination, gene duplication, gene conversion, chromosome rearrangements, horizon gene transfer and transposons.

the mitochondrion's close relatives, the *Rickettsia* bacteria,[12] which have 1,100 to 1,400 genes), it would not be an overstatement to describe this genetic exodus as a stampede.

How do we know there used to be more genes in the mitochondrion? Well, unlike a tidy renter who cleans up before moving out, genes are messy – they often leave bits and pieces behind. And that's how we know where they once were.* The mitochondrial genome now looks like a wasteland, littered with the remnants of those escaped genes. Without many of the essential genes it once had, the mitochondrion is trapped inside the cell, living a dependent life by outsourcing many of its crucial functions to the genes in the nucleus.

We've learned that the effect is limited by having small genomes, so what else can our struggling asexuals do to avoid Muller's ratchet? One effective strategy is horizontal gene transfer, where they exchange genetic material with their neighbours or pick up bits of DNA from their environment.† Horizontal gene transfer introduces fresh genetic material, saving asexual lineages from the relentless grip of Muller's ratchet.

Bacteria are masters of horizontal gene transfer, using three main tricks: transduction, conjugation and transformation.[13] Transduction refers to bacteria gaining genetic material from activities of viruses called phages. Conjugation involves the direct exchange of DNA between two individual bacteria. Often dubbed 'bacterial sex', this process has a whiff of

---

* Nuclear genomes often contain DNA fragments of mitochondrial origin, termed 'nuclear mitochondrial DNA' (or NUMTs). These DNA sequences also serve as evidence of past transfers between mitochondrial and nucleic genes.
† 'Horizontal' in this context contrasts with 'vertical', which refers to the better-known process of genes being passed down from parent to child.

A Ratchet, a Curse and a Story of Emergence

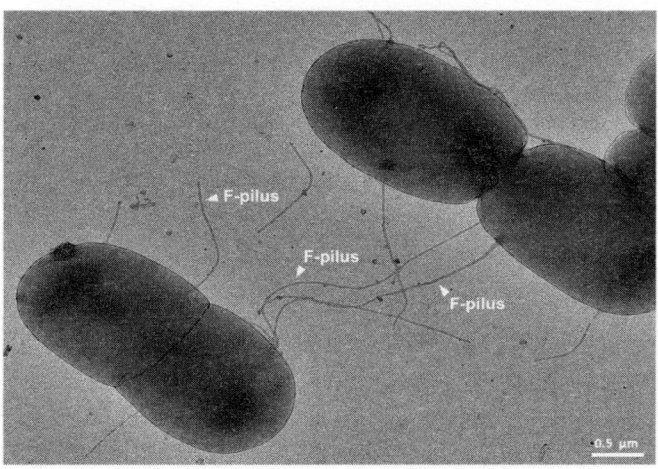

Figure 1.3. Conjugation in bacteria *E. coli*.
Note the thin tubes used for exchanging DNA –
marked as 'F-pilus' – between bacterial cells

intimacy: two bacteria connect, forming a small tube between them, allowing plasmids – tiny, circular DNA molecules – to pass through. Finally, transformation refers to a process where bacteria scoop up stray DNA fragments from their surroundings – like tiny genetic scavengers rummaging through nature's recycling bin, picking up garbage.

Picking up garbage may seem mundane to us, but in evolution the genetic equivalent of garbage-picking (that is, taking in DNA fragments from the environment) can be a matter of life and death in the face of the oppressive throttle of Muller's ratchet. Could more sophisticated eukaryotic organisms – protists, fungi, plants and animals – resort to such genetic manoeuvring as well, though?

This was long believed to be unlikely. However, bdelloid rotifers, a group of tiny aquatic animals, were the first to raise suspicions that more complex organisms might engage in horizontal gene transfer through genetic scavenging.

There are about 450 species of these rotifers,* which resemble leeches (or *bdella* in Greek) under the microscope. Genetic studies show that the last time they had sex was roughly 40 million years ago. Since then, these rotifers have lived a life of virtually eternal celibacy.[14] No wonder, despite being closely watched by biologists for three centuries, they have never been seen engaging in sexual reproduction.†

John Maynard Smith, who coined the aforementioned term 'the twofold cost of sex', was so frustrated by these creatures that he called them an 'evolutionary scandal',[15] as they seemed to defy all our theories about the origin of sex. Could they be an escapee from Muller's ratchet? If so, how do they manage it?

As it turns out, bdelloid rotifers, like bacteria, resist Muller's ratchet by absorbing bits of bacterial, fungal and plant DNA from the water they live in and incorporating it into their genomes.[16] This secret, uncovered by researchers with advanced molecular technology, revealed that these little creatures have been using this genetic trick for tens of millions of years without having sex, all while hiding amid pond scum.

More recent studies show that bdelloid rotifers aren't unique in executing this process either. Many types of microscopic eukaryotes also pick up genetic fragments occasionally.[17] You might wonder why few complex organisms rely on this genetic scavenging to diversify their DNA. One reason is

---

* Bdelloid rotifers typically possess relatively large genomes, containing approximately 30,000 to 40,000 protein-coding genes – substantially more than the estimated 20,000 protein-coding genes in the human genome.
† A recent study shows that some species of bdelloid rotifers may occasionally engage in sexual reproduction. (Vakhrusheva, O.A., Mnatsakanova, E.A., Galimov, Y.R., et al., 2020. Genomic signatures of recombination in a natural population of the bdelloid rotifer *Adineta vaga. Nature Communications*, 11(1), p. 6421.)

A Ratchet, a Curse and a Story of Emergence

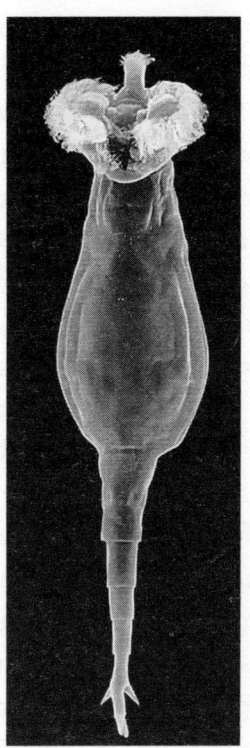

Figure 1.4. A scanning electron micrograph of a bdelloid rotifer, famous for exclusive asexual reproduction

that being too receptive to new genetic material can make a genome too unstable to function well.

Bacterial genomes, despite being quite open to changes, have evolved ways to manage this. For instance, they might use transformation to fix faulty mutations, protecting their genetic integrity from potential disruptions by foreign DNA. So, even though bacteria seem genetically promiscuous, exchanging genetic materials with others all the time, they appear to practise a form of 'safe sex' of their own.[18] Probably because of this, genetic intercourse in bacteria may not be suited for eukaryotic organisms, at least not in a major way.

The strategies above – transduction, conjugation and transformation – are some of the primary ways most asexual organisms (especially bacteria) dodge the relentless curse of Muller's ratchet.* However, most complex organisms aren't asexual, and the reason is simple: sex doesn't just allow organisms to duck Muller's ratchet; it also brings a whole host of perks. For starters, sex reshuffles the genetic deck, mixing genes from the broader pool and dealing out fresh combinations with every new generation. This genetic remix lightens the load of harmful mutations and helps keep Muller's ratchet from grinding things down.

But sex isn't just about cleaning up genetic messes – it's a crossroads where different evolutionary paths meet, like highways merging traffic from all directions. By fusing genes from diverse sources, sex equips organisms to adapt more quickly to their environments. These benefits make it clear why sex didn't just pop up – it's stuck around for the long haul. But hold on, we're getting ahead of ourselves.

You might have spotted a potential issue here. While shuffling DNA can bring beneficial genes together and speed up adaptation, can it also disrupt a set of good genes that are already working well together? As it turns out, yes. Evolutionary biologist Sarah (Sally) Otto likens it to a game of poker: when you're holding a strong hand, like a straight

---

* Nature may have other methods to fight Muller's ratchet. For example, scientists have discovered that segments of unused chromosomes (also known as B chromosomes) can be reserved for that purpose in a parthenogenic Amazon molly. (Schartl, M., Nanda, I., Schlupp, I., Wilde, B., Epplen, J.T., Schmid, M. and Parzefall, J., 1995. Incorporation of subgenomic amounts of DNA as compensation for mutational load in a gynogenetic fish. *Nature*, 373(6509), pp. 68–71.)

flush, shuffling and exchanging cards with others could end up weakening your winning combination.[19] So should organisms stop or slow down having sex once they've fully adapted to their environment? According to studies on creatures like rotifers, the answer is yes.[20]

Sexual reproduction isn't perfect. Asexual reproduction, though not without its flaws, has its advantages, as we have established. It's faster, allowing for rapid habitat takeover when conditions are favourable. Plus, in remote or isolated areas where finding a mate is tough, asexual reproduction is a lifesaver, offering a way to reproduce when sex isn't an option.

A key thing to note is that the perks of asexual reproduction are mostly immediate. On the flip side, the advantages of sexual reproduction – like countering Muller's ratchet and mixing in beneficial genes – often pay off in the long run. That's why asexual reproduction is so tempting for sexuals: it offers immediate rewards. Sexuals, meanwhile, have to constantly face the tricky task of balancing their long-term survival with the immediate gain in reproductive efficiency that asexual reproduction promises.

How can sexual species resist the lure of instant gratification from asexual reproduction, given its tempting short-term benefits? The truth is, they can't – because they don't have the luxury of willpower. They only respond to instant rewards and 'learn' the hard way through natural selection: once they veer into the fast lane of asexual reproduction, they eventually find themselves trapped. They may enjoy a brief burst of high-speed joyride success, but the only exit from that road is extinction – sooner or later.

This is true for many eukaryotic life forms, which include all complex, multicellular organisms. Occasionally, some sexuals do indeed abandon their biological identities due to the low efficiency of sexual reproduction and switch to asexual reproduction.[21] Yet in their rush for quick success, these asexual converts meet the same fate as countless life forms before them, doomed by Muller's ratchet.

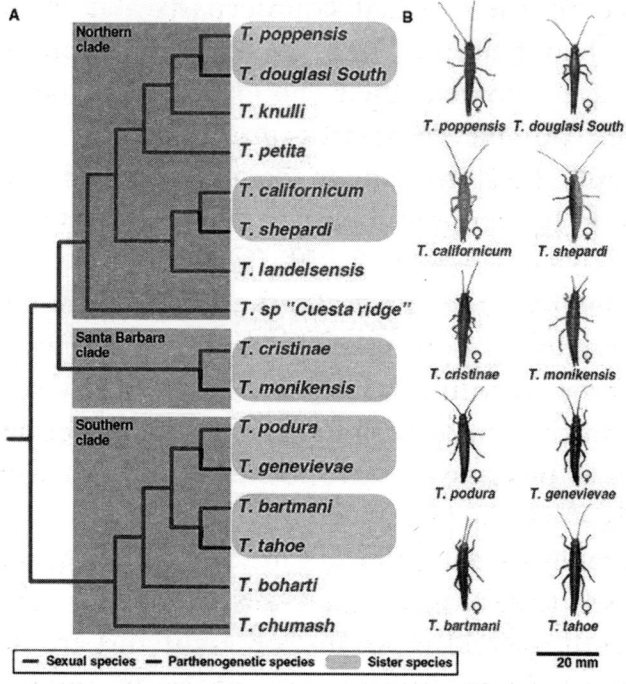

Figure 1.5. Multiple, independent transitions from sexual to asexual reproduction in a group of stick insects (genus *Timema*). (A) A phylogenetic tree showing the evolutionary relationships of the *Timema* species. For each of the five pairs of species (whose names are boxed within a shade), the top is sexual and the bottom is asexual. These five pairs of species are again shown in (B): sexual species are on the left, compared with their asexual relative on the right

Thus asexual efficiency, while alluring in the short term, may ultimately lead a species to an evolutionary dead end.

A Ratchet, a Curse and a Story of Emergence

That's perhaps why asexuals are often sporadically scattered amid sexuals across evolutionary trees (also known as lineages or clades) of life forms, including plants, animals and fungi. Their species longevities tend to be short, as shown by the short lineage branches in Figure 1.5 for stick insects, compared to their sexual counterparts.[22] If these asexual converts had been warned, 'Wrong direction, pal!' they might have avoided their boom-and-bust fate in their misguided pursuit of instant gratification. Unfortunately, nature only sees immediate benefits. Sexual species will continue to be tempted to switch to asexual reproduction.

We can see that both sexual and asexual reproduction have their pros and cons. Asexuals excel in reproductive efficiency but fall prey to Muller's ratchet, while sexuals can counter Muller's ratchet but can't match asexuals in reproductive efficiency. Is it possible to have the best of both worlds? The answer is a cautious yes.

This takes us back to California condors. Although they stick mostly to sex for reproduction, every now and then they dabble with asexual reproduction, as if just to keep things interesting.* Some species flip the script more radically: they're mostly asexual but throw in the occasional sexual fling.† Apparently, this switch-up gives them the best of both

---

* The evolutionary advantage of occasional asexual reproduction in predominantly sexual species is still unclear.
† An intriguing but lesser-known strategy is hermaphroditism. About 95 per cent of flowering plants use this approach. Close to 30 per cent of animals, especially invertebrates – like corals, flatworms, snails, earthworms, leeches, barnacles and starfish – are also hermaphrodites at least part of the time. Hermaphroditic individuals have both male and female reproductive organs, allowing them to reproduce on their own (selfing) or by mating and exchanging genetic material with others (outcrossing).

worlds – sky-high Darwinian fitness in the short term and a sneaky way to dodge the genetic dead end of Muller's ratchet in the long haul.[23]

Many plants, fungi and small animals employ this 'reproductive flip-flopping' strategy. Some take it even further by increasing their rates of DNA recombination during sexual reproduction, speeding up gene-reshuffling.[24] These are evolved measures that can reduce the frequency of sex needed for survival, thus minimizing its major drawback: reproductive inefficiency.

Among such reproductive flip-floppers are many species of plants like clovers, strawberries, bamboos and aspens. They too reproduce mostly asexually by sending out runners or stolons but occasionally engage in sexual reproduction.[25] Flatworms, nematodes, aphids and water fleas are examples of animals that use this trick.[26] They can detect environmental conditions like amount of light, temperature or crowdedness to swap between sexual and asexual states. This condition-dependent sex allows genes to switch to a better genetic community for survival, for which it's known as the abandon-the-ship principle proposed by Sally Otto and her colleagues Aneil Agrawal and Lilach Hadany.[27]

A prime example of this in action can be found in the water flea, *Daphnia magna*. They reproduce asexually when conditions are favourable. But when hardship looms, they shift to sex, producing males during droughts, famines, cold spells or the waning light of approaching winter. They also create a tough outer shell around their fertilized eggs to ensure survival through the cold and harsh winter months.[28] When favourable conditions return, they switch back to cloning again for high reproductive efficiency.

A Ratchet, a Curse and a Story of Emergence

Reproductive flip-floppers are a rare bunch in the vertebrate world. Apparently, it's not easy to shuffle between asexual and sexual modes when you've got a more complex body with specialized parts. Still, a few birds, snakes, lizards and even sharks manage to pull it off, sort of. The whiptail lizard, for example – one of the weirdest creatures out there – has almost perfected the switch from a wild night out to, well, cloning itself at home.

This creature can be found all over the American West, usually minding its business and reproducing the old-fashioned way – by sex. But in New Mexico and some nearby areas, there's a population that's ditched the dating scene altogether and gone fully asexual.

This switch goes back to a wild historical hook-up that took place as recently as a few hundred to a few thousand years ago. Two species, the little striped whiptail and the tiger whiptail, decided to mix things up and hybridize. This minor rendezvous led to a major reproductive roadblock: their chromosomes couldn't get in sync to produce sperm and eggs. Normally, it would be game over for the hybrids – think mules. But these lizards had an ace up their sleeve: they could reproduce without a partner through parthenogenesis, meaning reproduction by females alone. For this reason, they become a unisexual species of their own: the New Mexico whiptail lizard. How do they do it?

Since sperm cells can't make it on their own – let alone grow into whole lizards – male hybrids are a no-go. Eggs, on the other hand, are more self-sufficient but still need a spark to kick-start life in an all-female lizard community. So these crafty female lizards turn to parthenogenesis. They even mount each other to get things rolling, which triggers egg-laying without any fertilization in a process known as 'false mating'.*[29]

You might be wondering how these all-female lizards dodge Muller's ratchet, the genetic dead-end trap. As it turns out, parthenogenesis in these lizards is no simple copy-paste job like cloning, and is a bit more elaborate than creating identical copies, like Dolly the sheep.† Here's the deal: before they make eggs, female lizards double their chromosomes from diploid ($2n$) to tetraploid ($4n$), which then become diploid ($2n$) again after meiosis. Remarkably, they improvise some clever manoeuvres – crossing-over and pairing of sister chromosomes during meiosis.[30] This allows them to maximally preserve the existing genetic diversity they've got. Still, parthenogenesis doesn't whip up new genetic diversity, so the lizard can only slow down – but not dodge – the curse of Muller's ratchet. Here, as we can see, the big downside of being an all-female, unisexual species is the lack of a way to let in fresh genes. The New Mexico lizard seems to be doing just fine – for now, that is, and that's likely because they've only been around for a short while.[31]

If parthenogenetic species have taught us anything, it's this: Muller's ratchet might be mortal in the long run, but it's not strong or fast enough to beat the immediate perks of asexual reproduction. So odds are that it wasn't the thing that kick-started sex in the first place. Evolution is typically no fan of delayed gratification – it smiles on creatures that snatch quick

---

* Some parthenogenic species such as Amazon mollies and mole salamanders have a sneaky twist – they mate with males (even of different species in some cases), but not to fertilize their eggs. Instead, they use the sperm purely to kick-start egg development.
† The first mammal that was artificially cloned from a body cell.

rewards without a second thought. But Muller's ratchet is more like a slow-acting poison – it creeps up gradually, rearing its head only after many generations. That's hardly the kind of instant pay-off that would make sex so alluring. Even species that reproduce sexually can slip back into asexuality (just ask stick insects) when efficiency calls. So what's the quick fix, the immediate benefit, that made sexual reproduction a winning strategy from the get-go?

Many evolutionary biologists think the real answer still lies in mutation – but not the slow, creeping problem of Muller's ratchet. We're talking about the more dramatic, occasional mishaps: major DNA damages.* If left unrepaired, they can be fatal, especially when the population is small and at the risk of extinction. And when there's a break in both strands of that famous double helix, there's really only one way to patch things up: genetic recombination during sexual reproduction. In fact, from viruses to complex organisms, DNA repair happens everywhere, making sex often a necessity.[32] Best of all, this little trick brings both instant rewards and long-term benefits – just the kind of deal evolution loves.

Though compelling, the DNA repair hypothesis for the origin of sex still leaves us scratching our heads over one key issue: how much sex is actually necessary? Why is it so prevalent in the plant and animal kingdoms? After weighing the pros and cons of sexual versus asexual reproduction in this chapter, you might think the perfect strategy would be a mix of mostly cloning with the occasional romp to keep

---

* DNA damages can be caused by internal (such as errors in replication) and external factors (such as radiation, reactive oxygen species and viral infections). Small, minor DNA damages may be seen as part of genetic load that leads to Muller's ratchet.

Muller's ratchet in check, fix up any DNA glitches and reproduce efficiently. Sounds like a solid plan, right? But as usual, nature refuses to play along with our tidy theories. Some species seem to need a lot more sex than others, yet many others are at it almost non-stop. So why the difference? To get to the bottom of these tricky questions we need to dive head-first into one of evolution's most intriguing ideas: the Red Queen hypothesis.

2

RUNNING WITH THE RED QUEEN
The Maintenance of Sex

Most of us have dabbled in a little gambling, whether it's to add some drama to a sports game or just to win a cheeky bet with friends. For me, my most memorable brush with gambling happened in a casino, and it was momentous because it led to an epiphany about the evolution of sex. Yes, in a casino. Intrigued? Let me take you back to the spring of 1998 for the story.

It was my first time visiting 'Sin City' and, as expected, there were slot machines everywhere, even at the city's then McCarran airport. One evening, I decided to give gambling a go after dinner at Stardust. With a handful of quarters, I approached a nearby slot machine. On my twelfth spin, all three reels stopped at '7' – a jackpot of $300! The room erupted in excitement, lights flashing, chimes sounding, bystanders coming. Players near me dropped their levers, turning to watch the coins raining down into the payout tray.

My five minutes of fame left me with two heavy buckets of quarters. I instantly began to play again. It was basically free money at this point, and who knew, I might win even more. But when I left the casino hours later, I had emptied both buckets. Worse, even the ten dollars I'd started out with had been swallowed. So came the karma, as I'd been warned by the city's slogan: 'What happens in Vegas stays in Vegas.'

My experience was hardly special. Slot machines are designed to keep you on a short leash, with payout ratios ranging from 75 to 98 per cent. In plain terms, for every dollar you feed into the machine, you can expect to see 75 to 98 cents come trickling back. Of course, a few lucky spins might let chance have its say – anything can happen. I hit the jackpot on my twelfth try, out of pure luck. But when I kept going, the odds showed their true colours.* That built-in maths – one dollar in, 75 to 98 cents out – started to grind me down. Losing was inevitable.

Sex, in a sense, is like playing a slot machine – it is a thrilling gamble fuelled by the glittering promise of genetic jackpots. But when it comes to efficiency – the pay-off or the reward – sexual reproduction gets trounced by its no-frills rival, asexual reproduction. What struck me as a major revelation then was that the payout ratio for sex, as we've understood so far in the book, would be magnitudes lower than the stingiest slot machine in Vegas.

This becomes apparent when we consider reproductive flip-floppers like nematodes, aphids and water fleas. For many of them, sex is a rare, once-in-a-blue-moon kind of event – maybe once a year, or once in 100 or even 1,000 generations. If there were an award for sex abstinence, budding yeast would claim the crown: they bother with sex only once every *10,000 generations!*[1] That kind of commitment-phobia makes the pay-off for sex look ridiculously low – think 1 per cent, 0.01 per cent, or less (if the reward were

---

* In mathematics this is known as Bernoulli's law of large numbers, credited to the Swiss mathematician Jakob Bernoulli.

higher, we'd expect them to engage in more sex). By comparison, even the stingiest slot machines in Vegas offer payout ratios of around 75 per cent. Next to those odds, evolution feels downright miserly.

If some creatures can get by just fine without much sex, maybe this whole DNA damage and mutation problem isn't as bad as we discussed. Yet too much sex can be overkill, like using a cannon to shoot down a mosquito – too costly for a tiny benefit. So if the overall impact of harmful mutations doesn't immediately spell doom, there should be less pressure for organisms to engage in frequent sex, a point made clear by evolutionary biologist Hanna Kokko and others.[2]

But then we run into a paradox – a great one. Despite the apparent aversion to sex among many reproductive flip-floppers, most complex organisms still practise sex far more often.[3] In nearly all vertebrates, in fact, sex is not even an option but an obligation: do it or face imminent extinction, as they lack the capacity for asexual reproduction at all. How can we explain this apparent contradiction?

The bottom line is clear: for sex to become a permanent fixture of life rather than an occasional fling, the pay-off must outweigh the cost. Yet the very existence of reproductive flip-floppers tells us something crucial. Fixing damaged DNA and, to a lesser extent, beating back Muller's ratchet may have sparked the origin of sex, but they're hardly strong enough incentives to keep it going full-time. In fact, once sex took root, those threats were demoted from existential crises to minor annoyances – more like a dripping faucet than a flood. Occasional sex is enough to keep them in check, as these part-time lovers show us. So why do so many species stick with sex exclusively? What makes it worth the constant cost?

Here we can see that *the origin of sex is one thing, but maintaining sex is another*. The latter requires something more rewarding than simply repairing DNA and staving off Muller's ratchet.[4] There must be a bigger pay-off – one that explains why most higher plants and animals have permanently abandoned asexual reproduction in favour of sexual reproduction. If we stick with the metaphor of sex as playing the casino machines, the payout ratios would need to exceed 100 per cent to prevent eventual genetic bankruptcy. So what is this elusive advantage of sex that we've so far been overlooking?

. . .

For years, scientists searched far and wide for the hidden reasons behind the uncounted pay-off of sex, but progress was slow until biologist Leigh van Valen accidentally stepped in.* A truly original and poetic thinker, van Valen dedicated his entire life to intellectual pursuits. He was said to have filled his University of Chicago office with as many as 30,000 books, making it resemble 'a dark labyrinth'. When the author Matt Ridley paid him a visit for an interview, he found 'an oldish man in a checked shirt, with a grey beard that is longer than God's', buried behind 'ziggurats of balance books and three-foot Babels of paper'.[5]

In 1973 van Valen published a paper titled 'A New Evolutionary Law', reporting a surprising pattern in the fossil record: across millions of years, the survival rates of organisms – ranging from single-celled protists to the massive mammals of the Pleistocene – traced a linear line downward, like a slide in a children's playground. There was

---

* Debate lingers over the impact of van Valen's work on the evolution of sex before his Red Queen hypothesis gained traction.

some variation among species, but all in all the trend meant that all species within these groups faced a roughly equal likelihood of extinction. Van Valen dubbed this phenomenon the 'law of constant extinction'.⁶

This discovery was shocking. The biological world, as we know it, is often characterized by Tennyson's poetic observation – 'nature, red in tooth and claw' – of survival struggles between competitors, between predators and prey, and between parasites and hosts. Each species is evolving to keep pace with its rivals and enemies in a perpetual arms race. But if extinction rates remain constant across species as van Valen showed us, it seems the whole struggle-for-survival trope, so highly regarded ever since Darwin's era, is futile – a Shakespearean 'much ado about nothing'. But, van Valen wondered, if struggles between species are as unproductive as indicated by the data, why then do we still see them everywhere?

His search for the answer brought to mind a fairy-tale character from his childhood: the Red Queen from Lewis Carroll's 1871 fantasy novel, *Through the Looking-Glass*.* In the storied Wonderland, the Red Queen tells Alice: 'Now, *here*, you see, it takes all the running you can do, to keep in the same place.' Put another way, if you want to actually go somewhere else, you must run faster.

Translating the fairy tale into biological terms, van Valen thought, if the rate of extinction remains constant, then 'no species can ever win'. That is, species cannot increase their 'speed' to prolong their existence because they have already reached their maximum capability in their struggle for survival. This idea inspired him to propose the Red Queen

---

* Lewis Carroll, born as Charles Dodgson, was also a mathematician at Oxford University. Some believe his stories are metaphors of maths problems.

hypothesis: evolution is a never-ending arms race where species have to hustle at full throttle just to keep up with their adversaries and stay in the game.[7]

At first glance, van Valen's research article might seem like an irrelevant sidekick in the quest to understand sex. The word 'sex' doesn't even make a cameo. But for biologists in the 1970s, deep in the weeds of untangling the paradox of sex, it was a revelation, because the paper spotlighted a crucial yet overlooked cast of characters – parasites.*

Parasites (and, more generally, pathogens) are a nightmare for every living thing. For us humans they can sneak into places you'd rather not think about – hair, ears, skin, gut, liver, even the brain and, yes, down there too. Animals have it even worse since they can't pop by the doctor's office for a quick fix.† Plants aren't off the hook either, and even bacteria have their own tormentors: bacteriophages (or phages for short), viruses that love nothing more than a bacterial buffet. Yet despite this microscopic mayhem, life soldiers on. The secret to this tenacity? You guessed it – sex.

Here's how the system works. Parasites are like bloodsuckers, adapting their lives to blend seamlessly with their hosts, akin to a key slipping into a lock. But the hosts aren't just passive victims – they churn out new genetic variants to fend off these freeloaders. This sparks a wild game of genetic leapfrog, each side relentlessly trying to outsmart the other. It's an endless evolutionary race, with neither side ever truly

---

* In *The Red Queen* (London: Penguin, 1993), author Matt Ridley credits evolutionary biologist Graham Bell for being the first to bring the Red Queen hypothesis to the attention of biologists who study the origin of sex.
† Primates have been observed self-medicating with medicinal plants to combat parasitic infections.

winning. In the end, both parasites and hosts earn their survival through this ceaseless struggle.

The master concept behind the host–parasite arms race is an evolutionary force called frequency-dependent selection, where being rare actually gives you the upper hand. In the endless tug-of-war between hosts and parasites, parasites focus on the easy targets: the common, well-known host types. But when a new genetic variant shows up, it starts off as a rare oddball. Parasites tend to ignore it at first – it's just not worth the effort compared to the usual buffet.[8] This gives the new host a chance to thrive without much trouble. But as it thrives and becomes more common it catches the parasites' attention. Soon more parasites adapt to exploit the new host, leading to the erosion of its early advantage. Eventually, the once-rare host loses its edge and pays the price for its own success.

Here, for both the host and the parasite, when a new genotype starts flourishing its perks begin to fizzle out, making way for the next hotshot. So both sides are stuck in a never-ending scramble to crank out new genotypes just to stay in the game – it's a genetic tug-of-war that never calls time out. Like in Alice's Wonderland, standing still isn't an option. This constant back-and-forth leads to the rise and fall of different genotypes, making sex essential to the wheels spinning in this endless survival showdown.

In contrast, the build-up of harmful mutations is a slow grind for asexuals, and DNA repair may not be a constant emergency. In some cases, like bacteria and bdelloid rotifers, full-blown sex isn't even part of the playbook. These creatures have other tricks up their sleeves, like conjugation and transformation – or, in our term, genetic garbage-picking – to keep their genomes afloat from mutation load. So the real perk of regular or permanent sex isn't about patching up DNA or

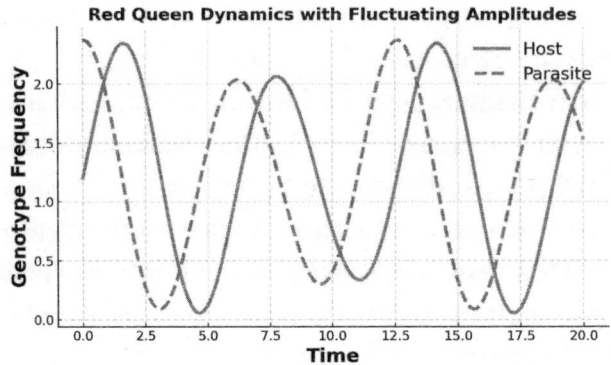

Figure 2.1. An illustration of hypothetical parasite–host interaction cycles. Shown here is only one of many possible genotypes in the host, and the matching genotype in the parasite. (Time unit is arbitrary: it depends on the species involved)

dodging Muller's ratchet, though this is why sex may have emerged in the first place. It's about staying one step ahead of parasites and pathogens by shuffling the genetic deck.

With parasites and pathogens constantly lurking, churning out clones with minimal genetic diversity just won't cut it. Regular sex has become a survival essential – a non-negotiable in this hostile world. We've now uncovered a key evolutionary force that makes sex not just a luxury but a necessity.

• • •

You might be wondering: does the Red Queen hypothesis hold up in the real world? Enter Curt Lively – a guy with the brains of a top-notch scientist and the muscles of a body-builder. To pay his way through college, Curt spent a year as an unskilled construction worker, hammering nails by day and catching waves by evening on the California coast. But let's be clear, his real love wasn't heavy lifting or hanging ten – it was science.[9]

The turning point came when Curt was admitted into the University of Arizona's ecology graduate program. He plunged into studying intertidal communities brimming with molluscs and barnacles. But before he could officially call himself a Ph.D. candidate, he had to face the dreaded qualifying exam – a rite of passage as nerve-wracking as it was necessary. This test, along with the equally harrowing thesis defence, was notorious for being about as enjoyable as a root canal. I could practically feel the stress, having once witnessed my own office mate, a fellow Ph.D. hopeful, pass out cold in front of a panel of no-nonsense scientists grilling him with merciless questions.

But Curt turned the stress into a spark. For his exam, one committee member handed him George Williams's thought-provoking book *Sex and Evolution* and asked him why the author believed that sexual reproduction posed a crisis for evolutionary biology, and what the solution might be. These questions were tough because there were no clear answers at the time. They were meant to test a candidate's depth of knowledge and capacity of reasoning. Curt was 'a little freaked out' at first. After months of reading and thinking, however, he not only passed the exam but also found himself deeply intrigued by the evolution of sex.

Life took a challenging turn for Curt post-Ph.D. In 1984, after successfully defending his dissertation, he faced a new hurdle. His wife, botanist Lynda Delph, secured a small stipend from the Fulbright Foundation for research at Canterbury University in Christchurch, New Zealand. Curt tagged along with neither pay nor job prospects. Arriving in Christchurch, they found themselves with only $12 to their names. To make ends meet, Lynda took up tutoring undergraduates in exchange for meals and a flat in a dorm on campus. Thankfully, a lifeline came from the New Zealand government, granting Curt a three-year fellowship to study the mud snail, *Potamopyrgus antipodarum*, a tiny freshwater species with a body length of only 4–6 millimetres (about 3/16th of an inch).* It was in this unsexy species that he found a perfect subject to study the evolution of sex, testing the Red Queen hypothesis.

Female snails of this species come in two flavours: asexual clones and sexuals who mate with males. Now, thanks to the cost of sex, we know that asexuals can reproduce faster than their sexual counterparts when all else is equal. So if we had a chance to pick at Curt's brain back then, we might have asked the same burning question: 'Why bother with sex at all?' After some early sampling, the penny dropped – parasites seemed to hold the key to the puzzle.

Snails are notorious hosts for parasites, especially trematodes – those nasty flatworms like blood and liver flukes that infect hundreds of millions of people worldwide. Once a mud snail gets hit by a trematode it becomes sterile, which means game over for reproduction. With the stakes this high, these snails are under serious pressure to fight back, and what better weapon in their arsenal against these parasitic invaders than sex?

Driven by this question, Curt hit the road in Lynda's clunking Volkswagen Beetle, collecting hundreds of snails from twenty-two lakes and twenty-nine streams. Back in the lab, he sliced them open and tallied the frequency of males at each site – males, of course, exist for nothing but sex, thus providing a direct insight into how frequently sex is occurring within a population. The more males, the more sex.

---

* It's an invasive species in Europe and North America.

Spotting male snails is pretty straightforward too, once you know the trick: just check for a penis behind the right tentacle. Once he had data on male population levels, Curt examined whether male presence was linked to the frequency of trematode infection in the host populations. If there was a positive relationship, he would be among the first to witness a signature of the Red Queen at work in the real world.

After several years of gathering, dissecting, counting and analysing, Curt's efforts paid off with a positive finding: in lakes where trematode parasites in the genus *Microphallus* were more prevalent, there were also more males present, indicating more sexual activity among the snails.[10] The study was repeated over a decade later, and the results still held up.[11] So, in conclusion, sex does react to parasites very well.

Beyond Curt's mud snails, the Red Queen seems to work across various species, with organisms large and small. Take the nematode worm *C. elegans*, for example. When infected with the bacterium *Serratia marcescens*, these worms show that mixing up their genes more by outcrossing than selfing (using its own sperm to fertilize its own eggs because some of the worms are hermaphrodites) is a winning strategy.[12] Flour beetles too up their genetic shuffling game when faced with their parasite, *Nosema whitei*.[13] Fruit flies likewise crank up their genetic recombination when battling infections from bacterial pathogens.[14]

In the wild populations of the gecko *Heteronotia binoei* in western Australia, parthenogenic individuals are practically mite hotels, hosting an average of 21.6 mites each. Meanwhile, their sexually reproducing cousins manage to keep their pest problems in check, averaging just 0.59 mites per individual.[15] Even dandelions get in on the action, switching between sexual and asexual reproduction in some populations. They stick to sex when pathogens like rust or pests like weevils are around, only reverting to asexual reproduction when the coast is clear with less threat.[16] In all these cases, as you can see, the Red Queen seems valid.

But here comes the downer: correlation is not causation. Just because two things seem connected doesn't mean one causes the other. For example, more ice cream sales might coincide with more pool drownings, but that doesn't mean ice cream causes accidents or vice versa. Similarly, married men live longer than unmarried men, but marriage itself might not be the reason, at least not the main one.*

In Curt's snail (and other species), finding more males in parasite-heavy lakes suggests a link between parasite infection and sexual activity, but it doesn't necessarily demonstrate that sex is the weapon hosts deploy against parasites. The next hurdle for the Red Queen – the hypothesized arms race between hosts and parasites – is to show that rare genetic variants outperform the common ones in fending off parasites.

But just as Curt was gearing up for this challenge, he hit a snag: Lynda's visa was about to expire. Luckily, she landed another postdoc at Rutgers, and once again Curt followed her.

In January 1989, when the couple left the sunny summer of Christchurch and their Kiwi friends, their feelings were tangled. I too spent seven months there in 1991, making many sweet memories: friendly people who waved at you from across the street, crisp, clear streams where ducks floated leisurely, and a charming university campus dotted

---

* The real reason for more drownings? More people are swimming on hot summer days. And as for why married men live longer, it's largely because women tend to marry men who are in good health and have stable finances.

with eucalyptus trees. Life was full of fun. Folks, even academics, held parties whenever they had a break – not just on weekends and holidays, but often during weekdays as well.

When Curt plunged back into the dead cold of Brunswick, New Jersey winters, he wasn't pleased. Living on Lynda's meager stipend, he once again found himself with no job nor prospects. His career was back to square one, just as it had been four years earlier when they first headed for New Zealand. What he didn't know was that this was a blessing in disguise. And, more interestingly, his path was about to intersect with that of Hermann Joseph Muller's.

As it turned out, several researchers at Rutgers were also hard at work cracking the enigma of sex, and after Curt gave a talk about his work with mud snails in New Zealand he and Lynda were invited to dine out at a local restaurant with two big-name biologists, Peter Morin and Bob Vrijenhoek, both of whom were Curt's academic heroes.

Science as a profession is a 24/7 job. Scientists often blur the lines between work and leisure. Instead of saying, 'How are you today?' they might greet each other with, 'How's the project going?' or 'Is your idea working?'. As you'd expect, the four of them chatted about their research at the dinner table. Impressed with Curt's talk, Bob brought up the topic of some little fish called topminnows.

Most topminnows reproduce sexually, but in the desert streams of Sonora in Mexico they share their habitat with a unique type of fish: a parthenogenic triploid ($3n$) fish, which has three sets of chromosomes in each cell, similar to an infertile, seedless watermelon. This triploid fish is a hybrid between two topminnow species, *Poeciliopsis monacha* and *Poeciliopsis lucida*. They reproduce asexually, just like the all-female whiptail lizard in New Mexico. Both sexual and asexual topminnow populations often suffer from black spot disease, caused by the larva of a trematode worm. These topminnows offered Curt yet another opportunity to study the relationship between sex and parasite load.

Bob had collected quite a few fish from Mexico over the years and stored them in the freezer in his lab. He asked Curt if he'd be interested in a postdoc position to work on the project.

'When do I start?' Curt was thrilled by the unexpected offer, feeling like he had found his lost career path.

'How about tomorrow?'

The very next day, Curt marched into Bob's lab. Partnering with Clark Craddock, a Ph.D. student under Bob's wing, he dived head-first into unravelling the mysteries of the evolution of sex, with a keen eye on the Red Queen hypothesis. Things started well, as expected: clonal fish were favoured targets for infections, while their sexual counterparts fared better. But then a curveball appeared – something Curt hadn't seen in his snails: when the sexual fish inbred, they lost their genetic mojo, becoming just as vulnerable to disease as the clonal fish.[17] Put differently, the power of sex against parasites comes from reshuffling the genetic deck – an advantage lost under inbreeding, which drains the deck of diversity.

. . .

The downsides of inbreeding shouldn't shock us. We've all heard of the monoculture mess in agriculture, which can only be blamed on the lack of genetic diversity. Take rice or bananas, for example – farmers pick a top-performing breed to boost yields, but that single-minded focus can backfire enormously when pests or diseases come knocking.[18]

When close relatives mate, it often exposes the otherwise hidden effects of harmful genes, particularly the recessive ones. The result is a host of problems: genetic diseases, reduced fertility, physical deformities and slower growth[19] (which is why, in most societies, it is illegal to marry a close relative). But there's more. Inbreeding can also turn a population into a magnet for pests, making them more vulnerable to parasites and pathogens and putting them at a higher risk of dying out.[20] This genetic turmoil, known as inbreeding depression, is a major concern for conservationists working to save endangered species like cheetahs, California condors and black-footed ferrets. These animals are already on shaky ground due to their small populations, and inbreeding only amplifies the risk, nudging them ever closer to the brink of extinction.

When small populations teeter on the brink, our primary concern isn't a slow genetic decline from Muller's ratchet but the sudden wipeout that a disease outbreak could bring.

Figure 2.2. The black-footed ferret was decimated by rabies and almost went extinct in the 1980s. Their strikingly similar facial features are probably due to a lack of genetic diversity rather than the fact that they're littermates

That's where genetic rescue comes in – a strategy to stave off extinction by introducing new genetic material, or, as they say, 'fresh blood', from outside the population. By adding genetic diversity, endangered species can bounce back with renewed vigour and a stronger defence against disease. This approach has already brought species like the Florida panthers and Illinois prairie chickens back from the edge.[21]

We can get a feel for how much of a difference genetic rescue can make by paying a visit to the wolf population of Isle Royale National Park, located on Lake Superior in North America. This population started in the 1950s when a few wolves crossed a temporary ice bridge to the island. With only about twenty-five individuals spread across three or four packs, they struggled with inbreeding depression on the 893-square-mile island. Their luck changed when a new male mysteriously arrived, bringing fresh genetic diversity and a brief boost in their health and vigour. Unfortunately, this improvement didn't last. By 2015, only two scrawny and sickly wolves remained: a male and his daughter (or possibly his sister), and neither seemed interested in mating.[22] To save the population, the National Park Service stepped in, spending $2m to reintroduce wolves from another population in 2018.* Thanks to this effort, the wolf population on Isle Royale persists today.

In contrast, inbreeding is rarely an issue for most small animals like nematodes, aphids and water fleas. These creatures usually have such massive populations that they over-

---

* This case reminds us how easy it is for earlier generations to wipe out a population, but how difficult – and expensive – it is for later generations to bring it back from the brink.

flow with genetic diversity. When times are good, many clone themselves so that they can rapidly expand their populations. But when the going gets tough – think cold snaps or dry spells – resources dry up and cloning just doesn't cut it any more. They switch gears from cloning or self-fertilizing to focusing on survival by resorting to sex. Sure, sexual reproduction slows down their baby boom, but it shuffles their genes enough to fight off parasites and pathogens. Plus it keeps Muller's ratchet, that genetic annoyance, in check, and could even repair damaged DNA if that's bugging them too.

Many of these reproductive flip-floppers slip into a cosy dormant state after a round of sexual reproduction, encasing their fertilized eggs in a thick, protective shell. Tucked snugly inside their little fortresses, the well-protected, fertilized eggs can ride out harsh conditions while letting the environment take care of their enemies.

This strategy gets even more dramatic with species like bdelloid rotifers, the genetic garbage-pickers we met in the previous chapter, who have yet another trick up their sleeve with which to fend off parasites. They can form tough, protective structures during harsh conditions, such as when their habitats – ephemeral puddles or shallow ponds – are parched. In this dormant stage their bodies dry out, which then kills off any parasites and pathogens. The rotifers are thus preserved the way dry grain is stored in a granary. When they rehydrate and become active again, often after being carried to a new location by wind, water or animals, they start anew without being bothered by their old enemies. This ability to evade parasites and pathogens allows them to sustain themselves with far less need of sex.[23]

Even more extreme than bdelloid rotifers are a group of eight-legged creatures known as tardigrades, or water bears, boasting around 1,300 known species. Some reproduce through parthenogenesis, while others use sex. Under normal circumstances, they're as vulnerable as any other aquatic invertebrate. But when tardigrades enter a dormant state, known as a tun, they become virtually invincible. Encased in a thick layer of body armour like crabs and lobsters, they can endure conditions that would be lethal to most other organisms. Even if you expose them to extreme temperatures, high pressures, oxygen deprivation, X-ray exposure, starvation or complete desiccation, chances are that you will fail to kill them.[24] In fact, they're so hardy that they are the only animals known to withstand the vacuum of outer space. This resilience allows them to thrive in environments where other animals can't. Needless to say, they can easily ditch their parasites by toughing it out through extreme conditions, which is perhaps why many of them can persist without sex.

· · ·

So we're back to that old debate: are reproductive flip-floppers the ultimate evolutionary winners? Well, it depends. These creatures can hit the gas when the environment's in their favour, rapidly swelling their ranks. But they can also face a nosedive when things go south. This rollercoaster lifestyle is what ecologists call r-strategy, where 'r' stands for reproductive rate. Think of mosquitoes or locusts – they're all about quick wins on high r values.

The r-strategist gambles on cranking out as many offspring and as fast as possible, without breaking a sweat over their future. It's all about quantity over quality – no time for parenting classes here. This short-term schemer thrives in tough, unpredictable environments. And when it comes to reproductive efficiency, who can beat asexual reproduction? That's why these reproductive flip-floppers often come out on top.

But in a stable or seasonal habitat, the future is more predictable. It rewards those who can plan ahead, rather than just living in the here and now. The opportunistic r-strategist, with its boom-and-bust lifestyle, is no longer the only player in town. Enter the K-strategist: rhinos, elephants and blue whales are some extreme examples.

While the r-strategist lives for the moment, the K-strategist has an eye on the future. It lives longer with lower mortality through good and bad times. It reproduces less often, with fewer offspring, but invests more resources and energy into each one, including parental care in some species, to ensure they grow up strong and competitive. It's a sturdy survivor, by fending off challenges from the physical environment and biological surroundings, including competitors, predators and, yes, parasites and pathogens.

The K-strategist opts for quality over quantity. Unlike the r-strategist, it doesn't go on a wild reproductive spree. Instead it keeps its population steady, around the environment's carrying capacity (that's the 'K', standing in for the 'C' in its name). The K-strategist plays the long game. And even with fewer offspring (with lower r values), it can still show some impressive growth over time. So while the K-strategist might not win the sprint in population expansion, it's built to win the marathon of survival and endurance.

With fecundity less of a concern, much in the cost of sex gets written off, making way for the rise of sexual reproduction among K-strategists. Meanwhile, size starts to matter. It helps you to outcompete rivals, evade predators and develop specialized body parts, like the immune system, to handle the omnipresent parasites and pathogens. These conditions all favour regular sex. That's why even the smallest vertebrates are magnitudes larger than nematodes, rotifers or water fleas, the common reproductive flip-floppers.

But here's the rub: imagine you're an animal. Being big comes with a hefty price tag – you need more food, more space, and a lot more of everything just to keep going. Your habitat starts feeling cramped fast, and your population shrinks compared to the swarms of microscopic creatures around you. That leaves you naturally more prone to inbreeding. You might still manage OK, but then – bam – a clever primate species called humans shows up. They hunt you, pollute your home and bulldoze your habitat, shrinking your numbers even further. Now, inbreeding isn't just a worry; it's a full-blown crisis. Look at today's endangered giants, typical of K-strategists: pandas, cheetahs, tigers, elephants, rhinos, whales and apes – all struggling with dwindling genetic diversity and shrinking family trees.

As we all know now, if pollution, over-hunting or habitat destruction reduces a population to the point where there's little genetic diversity left, sex will lose its mojo to create new genetic combinations to combat parasites, pathogens and bad mutations. Not only that but, as we've established, inbreeding can cause even more problems, leading the population to spiral down even further, beyond the point of no return.

This double-whammy of dwindling population size and vanishing genetic diversity is known as extinction vortex, or the Allee effect, a concept introduced in the 1930s by Warder Clyde Allee, an ecologist and van Valen's predecessor at the University of Chicago. The Allee effect works for a wide range of organisms,[25] possibly contributing to the demise of such iconic species as the passenger pigeon and the ivory-billed woodpecker. It shows that once a species enters this downward spiral, its extinction becomes inevitable without intervention.[26] The good news: even though the Red Queen still reigns supreme, successful efforts to save species like the Iberian lynx, the California condor and the black-footed ferret show that there is still a glimmer of hope for many critically endangered species to be rescued from the deadly vortex by human intervention.

• • •

Now, back to Curt. After five unsettled years on the fringes of academia, his pioneering work on the Red Queen – exploring the host–parasite dynamics in mud snails and the inbreeding effect in topminnows – finally earned him the respect he'd deserved. His dream career materialized in the winter of 1990, when he and Lynda moved to Bloomington to take joint positions as assistant professors at Indiana University, following in Muller's footsteps from forty-five years earlier.

With a fresh chapter in his career, Curt was free to revisit the questions he'd put on hold, testing the Red Queen. Every year, except during the COVID lockdown, he returned to New Zealand's lakes – both for research adventures and as a break from Bloomington's cold winters and his windowless basement lab. Plus, it was sweet payback for the summer he'd lost in the southern hemisphere when moving to New Jersey back in 1988.

One of Curt's top research priorities was to test whether rare genotypes outperform common ones – a key condition for the Red Queen hypothesis, as we discussed earlier. To crack this puzzle, Curt needed to uncover the snails' hidden genotypes, not just visible traits like body size or the presence of a penis.

Curt joined forces with Mark Dybdahl, a postdoc with a knack for molecular biology, and together they turned to allozyme electrophoresis – a protein-probing technique for uncovering the hidden genetic diversity in enzymes. What they found blew them away. They discovered that while sexual females are diploids ($2n$) that can mate with males, the asexual snails are triploids ($3n$), reproducing only by cloning themselves – just like the triploid topminnows he studied in Bob Vrijenhoek's lab. Even more unexpectedly, the snail clones were brimming with genetic diversity. Suddenly, Curt had the perfect set-up to test whether those rare genotypes could outperform the more common ones.[27]

Curt and Mark set out to test the hypothesis by comparing how snail clones fared against parasites. They collected clones from four lakes and found that in two of them, the clones with rare genotypes raced passed clones with more common genotypes in survival.[28] Although the results from the other two lakes were less conclusive, it was still a partial win for the Red Queen.*

This imperfect result led Curt to try a different approach. He and his team studied populations mixed with asexual (without males) and sexual (with males) snails and tracked their performance under parasite pressure. The findings were decisive: asexual clones, despite being faster in reproduction, suffered more from parasites over time, and some even went extinct. In contrast, the sexual populations remained stable and robust. Once again, sex proved to be a powerful defence against parasites by generating new genetic diversity.[29]

---

* According to the Red Queen hypothesis, common clones can be under-infected in some lakes and over-infected in others because parasites need time to adapt to changes in their hosts. This lag in the parasite–host arms race explains the variation, which, far from being a contradiction, actually supports the Red Queen's predictions.

You might be thinking, 'Wait a minute – the snail's lifespan is only about a year. What about animals that live longer?' More years mean more chances for parasites to latch on and wreak havoc. How do long-living species crank up their genetic defences? One logical answer: they stir the gene pot more when making babies. And guess what? That's exactly what happens in mammals: the longer a species sticks around, the more it shuffles its genes by undergoing higher rates of genetic recombination.[30]

The most direct evidence that sex helps in the fight against parasites and pathogens comes straight from the body's front-line defence: the immune system. Studies in several species have demonstrated that the genes that make immunoglobulins evolve faster than most other genes.[31] This breakneck speed keeps the immune system on its toes, always ready to tackle whatever new threats come its way.

Be aware that, although we've focused on the hosts, the benefits of sex in this arms race aren't one-sided. The Red Queen doesn't play favourites – she works for the parasites too. Genetic diversity should also help them stay in the game and fight for survival. Take the nematode *Strongyloides ratti*, for example. This roundworm parasite lives in rat guts and cranks up its sexual reproduction when the rodent's immune system puts on the pressure. In this evolutionary showdown, both sides are constantly shuffling their genetic decks to stay ahead, perpetuating the non-ending arms race.[32] Once again, the Red Queen hypothesis gets a thumbs up and so, both Alice and the Red Queen keep running.

As Curt, I and all other scientists have experienced, navigating the world of science is like battling in a gladiator pit, where ideas – any ideas, from those about DNA structure to climate change – are ruthlessly tested and only the strongest survive. For our intrepid Red Queen, it's no different – she's running a gauntlet of ever-tougher challenges.

Curt knew that to further challenge the Red Queen, more focused studies were required. But he never imagined that the spark for this endeavour would come from a surprising source – Jukka Jokela, a quick-thinking Finnish postdoc and trained SCUBA diver.

While collecting snails in a New Zealand lake, Jukka casually asked Curt if he'd ever explored beyond the lake's edge. 'No,' Curt admitted, a bit sheepishly. Without a word, Jukka grabbed a net, swam out, and began diving. After several rounds, he returned with some snails from 5 to 10 feet deep, living on *Isoetes* (mid water),* and others from 13 to 20 feet deep, holding on to the leaves and stems of *Elodea* (deep water).

The researchers pitted shallow-water snails against their mid-water and deep-water cousins. The results? Sexual snails were thriving in the parasite-infested shallows, where ducks – the fluke larvae's final hosts – loiter and unload parasite eggs by the thousands from their guts. Down in the deeper waters, however, it was a different story – here, asexual snails were having a population boom, with a mix of both types of reproducers appearing in the mid water. Therefore, the biggest parasite pressure was in the shallows, making it the ultimate selection battleground for these snails.[33]

But did the host and parasite engage with each other in a manner of cyclic dynamics? To find out, Curt teamed up

---

* Unlike the common *Elodea*, *Isoetes* – or quillworts – are tiny plants that reproduce via spores, much like ferns. My fascination with *Isoetes* started thanks to my son, Orien, who was just fourteen at the time. After two years of scouring, we finally struck gold in the knee-deep waters of Lake Sammamish, just east of Seattle.

with his Ph.D. student Britt Koskella for a multi-year experiment that put snails and parasites in a head-to-head evolutionary showdown. They infected asexual snails either simultaneously or with a one-generation delay, letting the hosts and parasites adapt to each other's moves. The result, once it came, was unmistakable: common clones became rare, and rare clones rose to prominence, with a one-generation lag showing that classic Red Queen dynamic. Additionally, the common clone was more resistant at the beginning of the experiment, but more susceptible at the end – a result also predicted by the Red Queen.[34]

• • •

Remember how we talked about sex moonlighting as a double agent in the host–parasite arms race? Could sex also play a key role in other interactions – not just between hosts and parasites, but between rivals, predators and prey, or plants and the herbivores? Van Valen himself believed that the nasty, cut-throat relationships – parasitism, competition, predation – pack a harder punch than the warm, fuzzy ones like mutualism and cooperation. So here's the real question: does the Red Queen's 'run just to stay in the same place' rule apply to these high-stakes battles, like the never-ending race between predator and prey? If that's the case, the rewards for frequent sex could skyrocket, making van Valen's Red Queen an even bigger evolutionary celebrity.

To answer this, a rigorous study is needed, something that can give us a large sample size in a relatively short period of time. In science, small is often beautiful, and Julia Haafke, Maria Chakra, and Lutz Becks at Germany's Max Planck Institute for Evolutionary Biology understood this perfectly

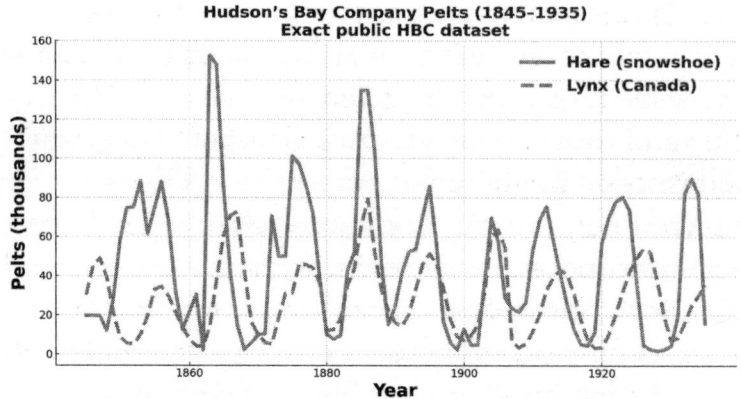

Figure 2.3. Predator–prey interactions such as those between the lynx and the snowshoe hare may also be subject to the Red Queen process

when they found their ideal study subject: a tiny species of rotifer called *Brachionus calyciflorus*.

This nifty little rotifer can reproduce either sexually or asexually, depending on its population density. It feeds on green algae, which comes in two types: one is single-celled and defenceless, easy pickings for the rotifer, and the other is colonial and defensive, not so easy to munch on. The researchers set up two experimental conditions: one where the predator–prey relationship stayed constant, and the other where they could evolve together in a classic arms race. The results? In the changing, arms-race environment, the rotifers showed a higher tendency to get frisky and opt for sexual reproduction.[35] This suggests that the evolutionary tango between predator and prey can indeed favour sexual reproduction.

Studies like this are eye-opening. They give us a clearer picture of the role of sex in the ecological world. As Haafke's team shows us, sex isn't some mystical enigma. Its primary job is to shuffle genetic decks, creating new combinations that help species survive in a cut-throat community where everyone's out to get each other. This suggests that other types of arms race – like those between cheats (mimicry, for instance) and those trying to catch them, or between plants and herbivores – might also drive up the need for sex.*

Unsurprisingly, we can see these same struggles playing out between antibiotics and antibiotic-resistant bacteria. Remember, bacterial 'sex' (formally, conjugation, as we mentioned in Chapter 1) is how bacteria share genes to gain antibiotic resistance. In all these hostile, antagonistic interactions, species are locked in an endless battle. One species makes a move, the other counters, and the cycle continues over time.[36]

At this point, we see that sex is way more than just a trick to dodge Muller's ratchet or repair damaged DNA. Think of genetic diversity as a stockpile of weapons, crucial for species to keep battling their foes. So the real magic of sex lies in its power to hand organisms fresh arsenal, ready to tackle tough challenges in the cut-throat world of ecological interactions. The Red Queen hypothesis, proposed half a century ago, seems to hit the bullseye when it comes to how sex is maintained, though not intended by van Valen himself.†

It's rather ironic that the Red Queen, a concept praised as 'one of the most important ideas in modern biology' by

---

* Sex is also widely believed to allow organisms to better adapt to unpredictable environments and gain the ability to exploit unused resources.
† Theoretical modelling with a broad range of ecological interaction between species shows that only strong selection may work. Even so, there are some exceptions. Thus the scope of the Red Queen in maintaining sex could be limited. (See Lively, C.M. and Howard, R.S., 1994. Selection by parasites for clonal diversity and mixed mating. *Philosophical Transactions of the Royal Society of London. Series B: Biological Sciences*, 346(1317), pp. 271–81. Otto, S.P. and Nuismer, S.L. 2004. Species interactions and the evolution of sex. *Science*, 304, pp. 1018–20.)

paleontologist David Jablonski, almost didn't make it out of the gate. Despite its groundbreaking potential, van Valen's key paper was turned down by several scientific journals, including *Nature*. But van Valen wasn't one to take no for an answer. Instead of sulking, he rolled up his sleeves and started his own journal, *Evolutionary Theory*, giving the Red Queen the spotlight it deserved in its very first issue.

· · ·

Over the years, I've bumped into Curt at various Evolution Society meetings. When I ran into him and Lynda in Montreal during the society's 2024 conference, I couldn't resist bringing up the big question again: 'Can we say now that the origin of sex, the queen of evolutionary problems, has been largely resolved?'

He caught my reference right away – the growing evidence for the Red Queen. He paused, thinking it over, and then, with a hint of doubt, simply asked, 'Do you think so?' He wasn't strutting with confidence despite four decades of the relentless pursuit in his lab.

This reminded me of a conversation with Sally Otto, whom we met in the previous chapter. A Washington, D.C. native, Sally has always had a passion for maths and biology. When she enrolled in Stanford University, she planned to dive into genetic engineering, reasoning, 'because genetic engineers must use math'. She laughs now at the memory. 'Then I got to college and found out they use pipettes, not math.'

With a little nudge from her advisor, Sally went to talk with Marcus Feldman, a well-known mathematical biologist. He handed her a puzzle about a Texas clam shrimp, a 3-centimetre (1.2-inch)-long creature living in ephemeral pools and temporary ponds. The coolest thing about this little shrimp is that it has three genetic sexes – one male and two types of females.* Sally was instantly hooked by the mystery. 'The females can self-fertilize,' she explains. 'So, I asked, under what conditions are males maintained?'

She found her niche and never looked back. With her sharp mind and groundbreaking research, Sally became a leading light in theoretical biology, particularly in the evolution of sex and genetic recombination.

Unlike Curt, who thrives on getting his hands dirty, collecting empirical data to test a string of hypotheses, theoretical biologists are a different breed altogether. Sure, they might keep small creatures like bacteria or yeasts around to stay in touch with the biological world, but their true passion lies in the untrammelled realm of possibilities by crunching maths formulas and coding their way through complex models.

'With so many genes involved, biological evolution is incredibly complex,' Sally muses, her thoughts turning philosophical. To her, the question of why sex exists isn't just one puzzle – it's 'a bunch of different puzzles'. She finds it fascinating to uncover all these additional pathways. With the freedom to let her mind wander far and wide, she has developed a range of potential explanations for how sex could have emerged and evolved. And the Red Queen is just one of these possibilities.[37] That explains why Curt is still reluctant to say that the puzzle of 'why sex?' has been solved.

Until these alternative theories are fully explored, the origin and maintenance of sex will remain one of the most thrilling frontiers in evolutionary research. Just as our grasp of the subatomic world continues to deepen, our understanding

---

* As we will elaborate in Chapter 5, it may be more appropriate to refer to them as 'genders' rather than 'sexes'.

of sex – the queen of evolutionary puzzles – will only grow richer and more intricate. We're moving from the simple to the complex, from the intuitive to the truly mind-bending, constantly pushing the boundaries of what we know, stretching our quest for truth ever onward.

Now I can see why Curt, Sally and so many others have chased down the 'queen' with such passion and conviction. It's the beauty of science – there's always progress but never an end to exploration. The thrill of uncovering new insights while knowing that the horizon is always expanding is what makes the pursuit of science so endlessly gratifying and rewarding for the curious mind.

⋯

Although the exact origin and evolution of sex remain unsolved, we can take pride in how far we've come on this journey of discovery. After half a century of relentless pursuit and rigorous testing, van Valen's claim in the Red Queen hypothesis that 'no species can ever win' still holds true. But is that too bleak a perspective? Sure, the Red Queen dynamics have been an unforgiving force, eventually driving species to extinction. Yet they've also been the catalyst for evolution, constantly pushing species to adapt and, in the process, giving rise to the enormous diversity of life we see today.

Perhaps neither Curt nor Sally anticipated just how much value – both monetary and intellectual – would spring from their discoveries, whether they emerged from studying snails or crunching mathematical formulas. Today, the Red Queen's legacy has burst out of biology's borders, spilling with gusto into everything from weed management to cancer therapy, tech innovation, organizational behaviour and the fierce arena of business competition. It's as if the Red Queen's principles, honed in the relentless thrust-and-parry of evolutionary arms races, are tailor-made for outpacing rivals in every corner of modern life. Banks, investment firms – yes, even Wall Street – are sharpening their strategies in the Red Queen fashion to outfox competitors in markets that move at breakneck speed. And naturally, politics, sports, AI and robotics have jumped into the fray, sparking strategic leaps and breakthroughs at every turn. With each new layer of understanding, it seems inevitable that more fields will fall under the Red Queen's spell. I like to call this grand expansion 'the cultural Red Queen principle', an idea that's shot far past van Valen's wildest imaginings. Long live the Red Queen!

In this chapter we've peeled back the layers to discover that while DNA damage and Muller's ratchet might have ignited the spark of sexual reproduction in the first place, they don't quite explain why sex sticks around. Our quest for answers led us to the Red Queen hypothesis, which suggests that the real perk – or at least a major incentive – of sex might be its ability to fend off the relentless assault from parasites and pathogens.

But there's a further question that just won't quit: why do we almost always end up with two sexes – male and female – in the wild world of sexual reproduction? Is nature hiding something from us, some untapped potential beyond this tidy little binary? And the real kicker: how does all this business of sex fuel the riot of biodiversity we see around us? Buckle up, my dear reader – we're about to dig into these juicy mysteries in the next chapter.

3

MALE, FEMALE AND BEYOND

Sex and Diversity

If we could hop into a time machine and zip ourselves back 2 billion years,* we'd see many wild creatures that once roamed – or more accurately, floated in the waters of – the Earth. That may sound thrilling – and for a biologist like me, it certainly would be – but to most, these ancient forms of life might be a bit of a snooze-fest, for they were all single-celled organisms invisible to our eyes. Although some might be seen moving around by beating their cilia or flagella under a microscope, they lacked the size and range of variety to pique our interest. No cool kelps, no epic dinosaurs, no majestic tree ferns, and definitely no Archaeopteryx† – not yet.

It would be tempting to attribute this lack of life-form variety to it simply being too early. After all, it takes time to cook up the intricate, attention-grabbing creatures we know today. But life (that is, the existence of organisms) had already been around for nearly 2 billion years since it first showed up on Earth (about 3.8 billion years ago). The real hitch – and the part that fascinates me – is that these ancient

---

* Estimates for when sex emerged on Earth vary vastly, ranging from 1.2 to 2 billion years ago.
† A famous bird-like dinosaur.

trailblazers were all about cloning themselves. And that was their undoing. Without a way to stir up fresh genetic material except small-scale horizontal gene transfer, they hit an evolutionary wall, unable to break into the guild of the more complex life forms that come to mind today when we think about the ancient Earth.

As we now know, cloning is the express lane organisms should drive in if their goal is to multiply and spread. However, it's a double-edged sword. Sure, it's speedy and efficient, but it skips the all-important DNA shuffle. That means no way to patch up major DNA mishaps or sidestep the slow, inevitable grind of Muller's ratchet. Even worse, without sex to stir in fresh genetic variety, clones are like all-you-can-eat buffets for parasites and pathogens. So even though asexual reproduction might seem like a quick ticket to success, it's often just a fast track to extinction.* In the end, all that hyped-up efficiency of cloning turns out to be more sizzle than steak.

Nietzsche might have avoided the mental collapse that led to his death if he had embraced the same fierce 'will to power' that drives life to survive. Even with extinction hanging over their heads, no biological species throws in the towel without a fight. They battle threats with grit and creativity, reassuring us with a powerful mantra: life always finds a way. Faced with relentless pressures from parasites, pathogens and a cascade of mutations, life stumbled upon its ultimate weapon: sex.[†]

---

[*] For 2 billion years preceding the advent of sexual reproduction, clonal organisms likely maintained genetic diversity through mechanisms such as horizontal gene transfer.
[†] Sex makes both hosts and parasites evolve faster, as we discussed in the previous chapter.

Now back to 2 billion years ago: all the microscopic players were hustling for an edge, cranking out clones at breakneck speed. But then, something truly unexpected happened. Two protists – those single-celled pond-dwellers still lurking in water today – decided to shake things up. Instead of sticking to the cloning routine, they merged and swapped genetic material before creating offspring. At the time, it might've seemed like just a blip in Earth's long timeline, but in hindsight this was the first spark of a quiet revolution that would change everything.

This ancient 'sex revolution' may have been a lucky accident, but it was exactly the jolt life needed. By jump-starting genetic diversity, sex became a game-changer – a rejuvenating elixir for existence itself. It offered a much-needed escape from the genetic glitches that had haunted clones for aeons and injected fresh vitality by blending the best genes from different lineages, creating a 'greatest hits' mix for survival and adaptation. With the grand debut of sex, life on Earth was transformed for ever. What we can appreciate now is that sex, like shuffling cards before a new game, creates genetic diversity by mixing up genes from two parents. This gives the next generation a better chance to win in survival challenges. Like a card game, shuffling the deck is just the prelude – a set-up for a brand-new game waiting to unfold. In the same way, the radical gene mixing of sex sets the stage for a fresh round of life to begin. And in this whirlwind of chaos brought about by sex, an explosion of diversity emerges, spreading across every level imaginable.

First, sex gives rise to males and females, setting off a cascade of innovations through what we often amusingly dub 'the battle of the sexes'. This battle entails a struggle between two reproductive strategists, played out with polarized emotions in us (and probably to a lesser degree in some other brainy animals) – love and hate, fidelity and betrayal, yearning and dejection, passion and depression. It's a new kind of arms race, yet it's more than that. This is because, despite the clash of interests, males and females are in the same boat, sharing a genetic stake – creating healthy, robust offspring. Accordingly, their interaction is a blend of cooperation and conflict. When it comes to innovation, however, it's conflict that steals the spotlight. As we explored with van Valen's Red Queen, these tensions don't just stir the pot – they're a potent engine for diversity, pushing boundaries and sparking innovations.

In evolutionary biology, the battle of the sexes boils down to a seemingly trivial issue: who, the male or the female (or more precisely, the proto-male or the proto-female), invests more in gametes – the sex cells tasked with carrying their genetic heritage. In the early stages of sex, this was a minor issue, as gametes were probably alike in size and contained similar amounts of materials, a condition known as isogamy, which literally translates to 'equal marriage'.

But everything changed when one crafty protist found a 'smart' way to make more gametes by investing less in each one, promising more offspring and more copies of its genes. Other protists couldn't afford to sit back and do nothing; ignoring this trick would lead to evolutionary suicide. So the race for smaller, cheaper gametes began, with each trying to outdo the others by investing less material and energy. This shift from a bit of corner-cutting to full-scale free-riding gave rise to a class of cheating gametes, specialized in taking advantage of larger ones, thus ending the era of tranquillity and equality on the planet of isogamy.

We can imagine the process by putting ourselves into this evolutionary story. Picture yourself as a go-smaller gamete, trying to outsmart your rivals. The key? Find and join forces with a large gamete still floating around. By merging with it, you can tap into its rich material and energy reserves, boosting your offspring's chances of survival. This is crucial because your junior will have to venture out on its own from a single cell, the zygote. But there's a catch: all your go-smaller rivals are 'thinking' the same way. To stay ahead you need to up your game, maybe by enhancing your motility. But there's a challenge – your rivals are also using every tactic you can think of, making it a constant race to stay ahead.

Meanwhile, as more gametes jump on the bandwagon of going small, your advantage diminishes as competition intensifies. Worse still, the mortality of your offspring creeps up due to a lack of resources, a consequence of your corner-cutting tactic. This race eventually comes to a head: when a zygote lacks enough resources to sustain itself it becomes a jet without fuel, unable to take off in life. The free-riding strategy, if adopted by too many, loses its adaptive edge and becomes counterproductive, following a biological version of the economic law of diminishing return, known here as 'frequency-dependent selection'. Furthermore, when large gametes become so scarce that the difficulty of finding them offsets all the advantages of this corner-cutting, going small becomes maladaptive. In this situation, what can you do?

Aesop's fable 'The Ant and the Grasshopper', concerns a grasshopper who has whiled away the summer singing and dancing. Meanwhile, the ants used their time industriously storing up food. Being the ant is all about delayed gratification. You put the hard work in early, but then are able to cosy up in winter with a stash of food while the grasshopper faces the consequences of instant gratification: cold and hunger. The battle of the sexes works in much the same way. By putting a little extra material and energy into your gametes, you can stockpile resources to boost your offspring's vitality. Suddenly, going large isn't just an option – it's the better road to success. While the go-smallers scramble to find and compete for the shrinking pool of larger gametes, the go-largers sit pretty, reaping the rewards of their resource-packed offerings. In a sea of tiny, energy-pinching sex cells, quality now outshines quantity, giving the go-largers a decisive edge. The balance tilts in their favour, and the race for reproductive success shifts. This game of give and take continues until an equilibrium is finally reached.

As time passes, a minor size gap between go-smallers and go-largers can turn into an uncrossable divide. Today, go-smallers have evolved to be so compact that they contain only the essentials, mostly for a life of a few hours to a few days in mammals:* a complete genetic package and some mitochondria – biological batteries – to power the gamete's one-way journey to its destiny: to meet a go-larger for fertilization.† This tiny, short-living gamete is now called a sperm, and its creator, male.

On the other hand, go-largers have transformed into eggs, and their creator is accordingly named female. Eggs can grow so large that they alone have enough nutrients for the survival and growth of the next generation, starting a life from zygote to embryo and beyond.

To put the size difference between sperm and egg into perspective, let's consider some examples. In salmon, individual sperm cells are invisible to the naked eye, appearing only

---

* In bats, sperm may survive for months.
† In humans, there are about 75–100 mitochondria concentrated in the tail of a sperm compared with 100,000 mitochondria in an egg.

in milky clouds in the water, while individual eggs are sizeable enough to be used as fishing bait. In the ostrich, sperm size is no different from salmon, but an egg can weigh around 3 pounds, equivalent to about two dozen Grade A chicken eggs and could feed both you, my dear reader, and me for a day. And if the moa, a group of flightless birds that went extinct roughly 500 years ago, were still around, their eggs would weigh nearly 9 pounds each! Relative to its size, the kiwi lays the largest egg of all existing species, with a mass up to a quarter of its body.

While not as extreme as in salmon and ostriches, the distinction between sperm and egg in humans is still noteworthy. As in all sexually reproducing species, human sperm are abundant and cheap. A single ejaculation can unleash hundreds of millions of sperm cells. In contrast, human eggs are fewer by magnitudes, with only about 400 produced throughout a female's entire life. This difference in gamete size paves the way for the grand unveiling of a new evolutionary force – sexual selection – a captivating topic we'll dive into in the chapters ahead.

・・・

The division between male and female introduces another aspect of diversity related to sex: the split between the diploid body (with two sets of the genome, indicated as $2n$) and the haploid gamete (with just one set of the genome, denoted as n). In simpler terms, *sex brings in two distinct forms of existence* – the life of the body cells and the life of sex cells. Metaphorically speaking, sex gives life a split biological identity endowed with different missions.

Now, the issues: how are these two forms connected? And which one carries more significance? Or, more familiarly, is a chicken (body) a means for an egg (gamete) to produce another egg, or is it the other way round?

These are the questions I'd have loved to ask Darwin had I ever got a chance to sit together at a table with him. However, by the time I made it to his Down House in the summer of 2018, enjoying the sun on the patio while savouring a sandwich and tasty bowl of pea soup for my lunch, I had already missed the great man by more than 136 years. So I had to seek out the answers on my own.

My search led me straight to another nineteenth-century evolutionary biologist: August Weismann, a German who wrestled with this very question. A gifted pianist with a passion for Chopin, Weismann might have pursued a career in music if science hadn't stolen his heart. I felt an immediate kinship with him because we both adore nature's endless quirks and wonders. But he took that love to gloriously ridiculous extremes. The man was so engrossed in chasing insects that he was said to once have blundered into a pond, ending up drenched. My closest parallel – so far – was an unplanned dive into a lake to rescue a half-conscious, fully panicked beaver that had just escaped anesthesia mid-demonstration. It was a memorable fiasco – especially since I was in the middle of showing two dozen students how *not* to handle a beaver. Let's just say the lesson was … hands-on.

Apart from his sharp eye for nature's subtleties, Weismann dedicated his talents and life to meticulously crafted experiments, exploring the mysteries of reproduction, inheritance, development and evolution across animals as varied as hydras, insects and vertebrates. He was also the first to grasp the evolutionary role of sex as a generator of variation – a genetic marble block for natural selection to sculpt, shaped by each species' environment. But it was a particularly bold experiment that etched Weismann's name into fame. He chopped the tails off newborn mice as soon as their eyes opened – a feat he repeated on an impressive scale: 901 mice across five generations, from October 1887 to August 1888. Why would he do such a thing?

Make no mistake, Weismann wasn't some sort of mad or sadistic scientist. His experiment was intended to test a prevalent idea that captivated the attention of evolutionary biologists of his time: acquired inheritance. The theory – known as Lamarckism – proposed by the French naturalist Jean-Baptiste Lamarck in the eighteenth century, suggested that animals could pass on traits acquired during their lifetime to their offspring. According to this view, giraffes could make their children's necks longer by stretching their own necks. The same idea could also be used to explain the evolution of long legs in birds like stilts, herons and cranes. (Wouldn't it be nice if our parents could get plastic surgery and pass on those perfectly shaped noses to us?)

If so, Weismann reasoned and tested, removing a mouse's tail at birth would likewise lead to shorter tails in subsequent generations. The results, as he reported during a conference in Cologne in late September 1888, showed a resounding 'No!'

But Weismann was aware of a potential pitfall in his data: five generations might be too short for evolutionary changes to show up. So he decided to play the long game, extending his experiment for several more years, persisting to the nineteenth generation (those poor mice).[1] In this extended saga, he continued his tail-removing escapades with thousands of mouse pups. The outcome, unsurprisingly, remained the same. This experiment served as the proverbial nail in the coffin for Lamarckism, conclusively sealing its rejection within the scientific community.

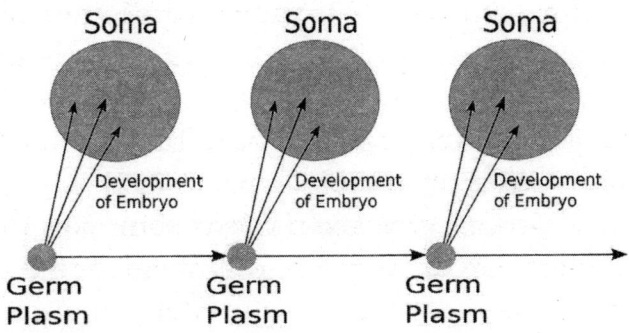

Figure 3.1. Weismann's theory about soma-germ division

From this study and many others, what Weismann figured out is that life operates with two separate cell lines: one for the body (soma) and one for reproduction (germ or germplasm that leads to the production of gametes). The germ line is immortal, preserving and relaying genetic information from one generation to the next, while the body is transient, playing no role in inheritance, as his mouse-tail experiment showed. The flow of genetic information, therefore, is a one-way street – from the germ line to the body (Figure 3.1), as if there existed a barrier to prevent backflow in the traffic.*

Despite his prolific academic output, Weismann never explicitly spelled out what the body does exactly. But if we read between the lines, his ideas suggest something striking: the body is merely a puppet, pulling whatever strings necessary to safeguard the survival and passage of the germ line.†

---

* This concept is known as the Weismann Barrier. From a modern scientific viewpoint, it's not entirely correct. At the very least, the body (soma) can profoundly influence the germ line through regulating the gene activities in the gametes.
† Weismann thought that germ-line cells were hidden somewhere in the body of animals without knowing that they are from somatic cells.

This is as close as we can get from Weismann for the answer to the age-old question of how the chicken and the egg are related.

Unfortunately, Weismann's research was largely done before the scientific community understood what genes are and how they work. As a result, his ideas about inheritance, development and evolution are quite limited from today's hindsight. This prevented him from putting the pieces of the puzzle together and finding the deep connection between the soma and the germplasm. To explore the territory that Weismann left uncharted, let's return to our notion of the two forms of existence – the diploid body and the haploid gamete – and see how sex vastly enriches the world.

In animals and seed plants, the diploid body is undeniably the boss – a commanding powerhouse of size and activity, orchestrating life's essential functions. The tiny haploid gamete, by contrast, plays a quiet role, tagging along without much fanfare. But in fungi and algae, interestingly, the script is flipped entirely. Here the haploid form takes centre stage, thriving as the star of the show and handling the critical tasks of survival. The diploid stage, meanwhile, is reduced to a fleeting cameo – a brief and often inconsequential appearance. This role reversal between the two forms of existence is remarkable, equivalent to your body shrinking to a single cell while your gamete takes control, transformed into the dominant survival machine, fully outfitted with all the 'bones and muscles' needed for life.

Zooming out to the big picture, the haploid and diploid stages in complex organisms take turns like playing a game of tag, creating a life cycle that spans generations. This is how sex adds a creative twist, tweaking the parameters of these two forms of existence. By favouring either haploids or diploids as the dominant survival form, it carves out countless evolutionary paths in terms of life cycle, fuelling the incredible biodiversity in life history we see today.

What does life look like when the haploid form takes the spotlight? Let's use fungi to get a feel. In mushrooms, for example, haploid individuals take the form of filament-like mycelia, which house gametes. A mycelium can receive nuclei (playing the proto-female role) or donate nuclei (playing the proto-male role). Consequently, many cells within a mycelium can give to or take nuclei from one or several genetically compatible mycelia.[2] If a mycelium cell gets a nucleus from another mycelium cell, the two nuclei often stay separate for quite a while – like newlyweds on a bizarre honeymoon, refusing to consummate their love in some kind of fungal chastity ritual.

Things get even more intriguing with yeasts, those single-celled wonders of the fungal world that give us nutrition and pleasure through the art of fermentation: bread, beer, wine, cheese, yogurt. Yeasts generate haploid sex cells of the same

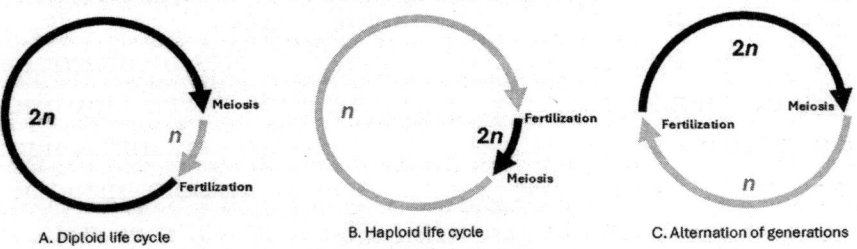

Figure 3.2. Simplified diagrams of three main types of life cycles. A. Diploid life cycle: dominated by the diploid stage ($2n$, black arrow). B. Haploid life cycle: dominated by the haploid stage (n, grey arrow). C. Alternation of generations: independent life form for both diploid and haploid stages

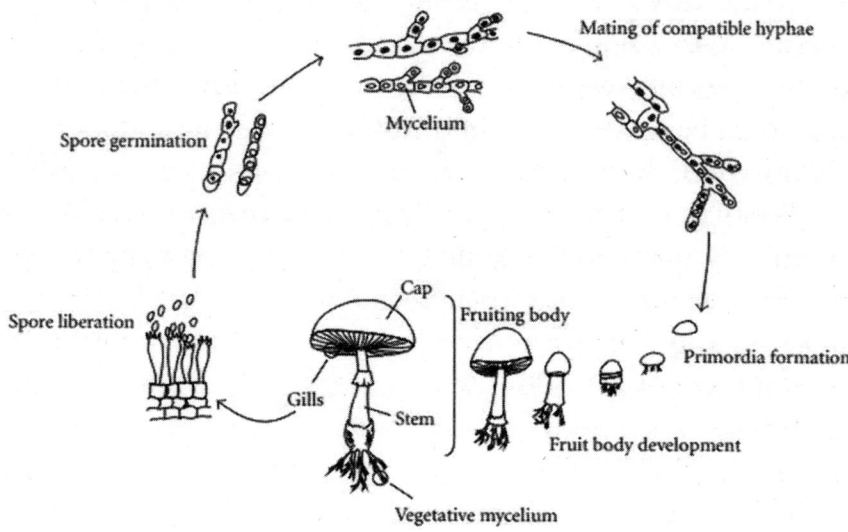

Figure 3.3. Mushroom life cycle

size (isogamy, which we talked about earlier) and mate by merging with other haploid cells. After fertilization the diploid cells quickly undergo meiosis, turning back into haploid cells, which then reproduce asexually. The fleeting diploid stage here seems to serve little more than a genetic mixer, shuffling genes from different haploids.

Are yeasts sexual or asexual? They're both, by all means, and alternate between the two in their life cycle. Are they reproductive flip-floppers? Sort of, but not exactly. In fact, they don't fit into any category we've seen before; they have carved out a path of their own. Here haploids dominate, in terms of size and control of living functions. And to borrow a saying – a chicken is an egg's way of making another egg – the short-lived, much-degenerated diploid becomes a haploid's way of producing another haploid in yeasts. (This stands in sharp contrast to the reproductive flip-floppers we've discussed, where the diploid form always takes centre stage.) With roles of diploids and haploids swapped, many biological rules we know – like the size difference between sperm and egg, the concept of male and female – don't apply. We might as well throw out the whole rulebook.

Peeking into the private lives of yeasts helps us understand why many fungi (and algae) prefer being haploid most of the time, unlike seed plants and animals. One big perk of this haploid-dominant lifestyle may lie in lowering the twofold cost of sex.[3] Since both parents (haploids) contribute equally to their offspring (also haploids), no 'parasitic' or 'useless' males exist. Everyone (a proto-male or a proto-female) lives, works and contributes like a typical female,[4] maintaining the efficiency of asexual reproduction while still generating genetic diversity through sex. This haploid dominance may stand for a smart strategy to balance and sidestep the costs of sexual reproduction.* It's a clever tack that opens up new survival opportunities and blazes new evolutionary trails, adding to the vast diversity of life on Earth.

The efficiency achieved by these haploid-dominant organisms is significant. In fact, the largest known single organism on Earth today is arguably a fungal species called *Armillaria ostoyae* in Oregon's Malheur National Forest. Its mycelia, the thread-like structures, spread in the soil, cover a ground area of 3.5 square miles. Remarkably, it has sustained and thrived in the same area for about 10,000 years, dating back to the

---

* For those who wonder whether there are transitional forms of life between diploid- and haploid-dominant species, the answer is yes. They are known as species which have the alternation of haploid and diploid in their life cycles. In plants such as liverworts and mosses, haploids are more developed whereas diploids mostly live on haploids. In ferns, however, both haploids and diploids are free-living.

Figure 3.4. Honey mushrooms (*Armillaria ostoyae*)

dawn of human agricultural civilization in Mesopotamia. The fruiting body of this organism is the honey mushroom, considered tasty by many mushroom enthusiasts.

• • •

We're all familiar with the two-sex system – it's the one we grew up with. But let's not kid ourselves into thinking it's the only game in town. Far from it! Nature, as always, loves to play by its own messy, rule-bending script, throwing in plenty of curveballs just to keep things interesting. And when it comes to sex, it's no exception.

The two-sex system dominates among multicellular organisms today, such as plants and animals, where males and females are distinguished by the size of their sex cells – a phenomenon known as anisogamy. In this system, biological sex is fundamentally defined by gamete size essential for *direct* reproduction: males produce the smaller gametes, sperm, while females produce the larger gametes, eggs.

Male, Female and Beyond

However, this traditional definition of biological sex, while widely accepted in biology, is not the full story; as we'll explore later, it fails to account for a rich spectrum of intersex variations shaped especially by kin selection.

But when it comes to single-celled organisms like protists, yeasts and slime moulds, things are different. They're isogamous, meaning their sex cells are the same size.[5] Unlike the anisogamous sperm cells, which can't survive long on their own, most isogamous sex cells have enough resources to live independently. Some, like yeasts, can even clone themselves. Since the sex cells are similar in size, we can't really call them male or female in the traditional sense. These little organisms only have one sex – they're unisexual.[6] But unlike some parthenogenic fishes and lizards that reproduce only by females, these unisexual organisms still engage in sex. How, then, does sexual reproduction take place without males and females?

The answer lies in mating types. Unisexual organisms engage in sexual reproduction between these types, which

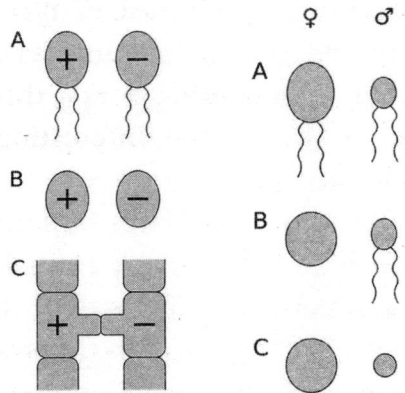

Figure 3.5. Isogamy in unisexual organisms (left), versus anisogamy (right)

Figure 3.6. The split-gill mushroom
(*Schizophyllum commune*)

are likely the predecessors of different sexes.[7] Despite lacking obvious distinctions, sex between mating types – call them proto-sexes, as we've already done earlier – functions just like all other forms of sex, reshuffling and diversifying the genetic make-up.[8]

There is one main difference, though. Unlike species with two sexes, unisexual species aren't confined to just two types of reproductive cells, like sperm or egg. On the contrary, mating types can go beyond – sometimes far beyond – binary. For example, in the slime mould (*Dictyostelium discoideum*), there are three mating types; in the ciliate (*Tetrahymena hyperangularis*), there are four; in the hairy protozoa *Tetrahymena thermophila* we met in the Introduction there are seven, and in the common mushroom (*Agaricus bisporus*) we buy from grocery stores there are eighteen. The fairy inkcap mushroom (*Coprinellus disseminatus*) boasts an impressive 143 mating types. However, the split-gill mushroom (*Schizophyllum* *commune*) takes the crown with a remarkable total of 23,328 mating types (Figure 3.6).*

How do we know that split-gill mushrooms have that many mating types? In the realm of mushroom romance, the prelude to sex involves the exchange of pheromone signals, crucial for sensing compatibility between potential mates.[9] This process involves two pairs of alleles (different versions of genes) on two different locations (loci), which determine mating types. For the split-gill mushroom, the first locus has 288 possible combinations, while the second has eighty-one. The total number of potential mating types is the product of these two numbers, resulting in a staggering 23,328.[10] If we equate mating type – or proto-sex – with sex, there would be that many distinct 'sexes' in this single species!

Picture the complexity of sex life for these organisms – if they had a dating app like Tinder, the algorithm would need to be far more sophisticated with a lot of 'if-then' logical switches. Luckily, reality paints a simpler picture. In most unisexual systems there's a straightforward rule: you can mate with individuals of any other mating type, just not your own. It's a bit like some traditional societies where people avoid marrying others from the same village if they share the same family name. Apparently, this 'marriage taboo' in unisexual organisms also acts as a check against inbreeding.

Even though there are many cases of multiple mating types in unisexual species, most of these creatures, like yeasts (*Saccharomyces*), ciliates (*Blepharisma*) and green algae

---

* Split-gill mushrooms are a common sight in North American woods, thriving in a variety of habitats – including, upsettingly, inside your nose.

(*Chlamydomonas*) still stick to having only two mating types.[11] Why is that?

As you've probably seen, biology often poses seemingly simple questions that prove challenging to answer. The dominance of the two-mating type system in the natural world is just such a puzzle. While there are a dozen or so hypotheses proposed, none has reached a conclusive status. One plausible idea posits that having two mating types helps signalling between mating parties, thus maximizing the chances of finding a compatible partner.[12] Although this logically makes sense, it still faces the challenge posed by some exceptions.[13]

In contrast, the prevalence of the two-sex system in anisogamous organisms, like animals, is relatively easier to explain: it stems from the fundamental differences between sperm and eggs.* As shown in the bonnethead shark, the Komodo dragon and the California condor, eggs are large and have the potential to brace for virgin birth, or parthenogenesis. Sperm cells, however, are tiny and carry minimal resources. Their brief lifespan doesn't allow for sustained existence, let alone developing into an embryo.

Think of males in the biological world as reproductive sidekicks or, more appropriately, parasites. Their entire existence revolves around passing on genes by finding eggs from females. Without sex, males would be pretty much useless. In fact, they can even make the whole process a bit inconvenient for females due to the cost of sex. That's why many species like snails, aphids, water fleas and social insects mostly do away with males by going solo – reproducing

---

* Hermaphrodites, when coexisting alongside males and females within a population, are sometimes regarded as constituting a distinct third sex. (See Lents, N.H., 2025. *The Sexual Evolution: How 500 Million Years of Sex, Gender, and Mating Shape Modern Relationships* (New York: HarperCollins).

asexually. Males pop up only when sex is absolutely needed. The rest of the time, they're like background characters or extras in films – few would pay them any mind.* So the popular punctuation play – 'a woman without her man is nothing' – has no ambiguity in biology: 'a male, without his female, is nothing'. But the fun part is yet to come.

Can sperm cells change course and become larger? If this were possible, it could give males a chance to redeem themselves. Unfortunately, the answer is no.† Why not? Take the human sperm cell as an example. It is 26,700 times smaller in volume than a human egg. Even were a male to double the size of his sperm cells, the sperm's contribution to an offspring's survival is still dwarfed by the egg's. And that size increase also comes with a serious, and possibly debilitating, trade-off – it slashes the total sperm count in half. Suddenly, this innovative male would be at a disadvantage in the race against any rivals.[14] The cost-to-benefit ratio is heavily against sperm as a result of rebelling from nature's plans. The takeaway: sperm are stuck being small, with no chance to turn around. Once a sperm, always a sperm.[15]

This logic applies to females too. The divisive selection force on gamete size is so powerful that once sperm and eggs have diverged to a certain extent, it's unlikely they can reverse course.[16] So no viable strategy exists beyond the bipolar choices of going small or going large for gametes. This

---

* Please don't read too much from biological facts presented with metaphors here. The meaning of human life should never be purely from scientific truths. Otherwise, we commit the naturalistic fallacy for the failure of making a distinction between 'what is' and 'what ought to be'.
† When gamete differentiation is small, there may be some flexibility for a back-and-forth switch between isogamy and anisogamy, as observed in certain protists. However, to date no such cases have been documented in animals.

explains why the prevailing system is based on two sexes. The situation is comparable to the electoral system in the United States. Unlike parliamentary systems in other democracies, the winner-takes-all set-up in America leaves little room for a third party to play a significant role in politics, though other factions do exist.

As we'll see in the next chapter, the idea that sex has to fit into a neat, binary box only holds up when it comes to direct reproduction. But alternative systems can also sustain when genes get passed along *indirectly*, in other words, when sex lets you spread your genes without having kids of your own. If you're confused by this proposition, you're in good company – I was too, back in college. It took me ages to wrap my head around the fact that evolution isn't about individuals; it's a game where genes call all the shots.

. . .

My love affair with biology began as a total accident – one I wasn't thrilled about at first. In 1982 I applied to study psychology at East China Normal University in Shanghai, but fate (and a less-than-stellar score on the gruelling, three-day National College Entrance Examination or *Gao Kao* in Chinese) had other plans. That test covered everything – Chinese, English, Maths, Physics, Chemistry, Biology and Politics – and I didn't exactly ace it. The university turned me down for psychology and stuck me in the Biology Department instead. I spent my entire freshman year sulking, convinced I'd been handed the wrong career script.

Then, during my sophomore year, something unexpected happened. While thumbing through a Chinese translation of Richard Dawkins's 1976 masterpiece *The Selfish Gene*, I was floored by the radical ideas in it. Dawkins claimed that genes are selfish, and that we – organisms – are nothing more than their vehicles, machines crafted for their survival and shuffling to the next generation. My reaction? 'What kind of nonsense is this?' I was determined to tear his argument to shreds.

Armed with nothing but my flimsy undergraduate knowledge, I waged days of mental warfare, attacking his ideas from every possible angle. But no matter how hard I prodded or interrogated, I couldn't poke a single hole. Begrudgingly, and with a good dose of lingering scepticism, I admitted that Dawkins might just be right. That reluctant acceptance marked a turning point. The upshot? I tumbled head-first into biology – and, by sheer dumb luck, ended up majoring in exactly the right thing.

I wasn't alone in my initial reaction. Many readers bristled at Dawkins's description of genes as selfish, thinking he was attributing emotions or intentions to them. The confusion, I later discovered, stemmed from a misunderstanding. Dawkins was talking about *teleonomy* (not to be mistaken as *teleology*, or purpose), evolutionary adaptation that appears to be purposeful: genes act *as if they were selfish* to outcompete others in evolution. The idea is that if they didn't behave this way, they'd get weeded out by natural selection.* Given the framework of teleonomy, anthropomorphizing genes with terms like 'cooperative', 'selfish', 'cheating', 'outlaw' and even 'killer' work surprisingly well and rarely lead to confusion once we jump this hurdle of miscommunication. These vivid descriptors have become part of the biologist's lexicon precisely because they capture complex biological

---

* The word 'selfish' could have been replaced by the more appropriate 'self-serving' in Dawkins's *The Selfish Gene*. But that would make the book title far less provocative.

phenomena in a way that's easy to grasp and, frankly, hard to resist, as a result of their relatability.

From this gene's-eye view, haploids and diploids are just two different vehicles for genes. In animals, these vehicles are highly specialized for their biological tasks. Haploids have one job – fertilization – while they outsource almost everything else to the diploids. Let's take a look at how evolution constructs Dawkins's vehicles for male and female bodies.

Since sperm cells are too tiny and ill-equipped to survive by themselves, they enlist support cells for nourishment, leading to the creation of a special tissue. To churn out more sperm, a part of the body takes on the job full-time – voilà, the testis is born. To ensure these little swimmers reach their destination, other traits have evolved too: muscles to propel, bones to support, hormones to regulate and a brain to strategize. Every part of the male body is geared towards one ultimate mission: making its sperm successful. Alongside these physical tweaks, a whole suite of physiological and behavioural traits has emerged for finding, wooing and mating with females.

Except for the ovary, the primary sex organ for producing eggs, the female body follows the same principle. However, there's a thematic variation: females tend to focus more on maximizing the survival of their eggs rather than just finding and mating with a partner, like males do.

This twist in the design of male and female bodies leads to what we call secondary sexual characteristics (in humans: genitalia, body hair, voice pitch, growth patterns, etc.) in some species. They are associated with sex but are not sex itself because they are not directly involved in producing gametes. (We will elaborate on this in Part II.) Surprisingly, secondary sexual characteristics, as a recent study in mice revealed, encompass a range of traits far beyond what we had assumed.[17] Their roles are broad but all focus on one goal: ensuring reproductive success. They aim to accomplish this by competing for, finding and attracting mates; producing, delivering and fertilizing reproductive cells; housing, nourishing and preserving embryos; and nursing, protecting and raising newborns.

In most animals, the diploid body is so meticulously engineered and fully equipped that it leaves the haploid gamete practically powerless. Picture the diploid body as a fully loaded UPS truck, while the haploid gamete is just the package – precious cargo, yes, but utterly dependent on its transport. This perspective pushes beyond Weismann's soma–germplasm (also cast as the chicken–egg) divide: neither form exists for itself, but rather for the genes it ferries forward.

• • •

The concept of the selfish gene also highlights another remarkable aspect of sex: by breaking up a clonal genome and putting genes in different communities, *sex bestows genes individuality*. This change is revolutionary as it sets in motion a cascade of processes that leads to more diversity and complexity, extending from genes and individuals to social groups, populations and species.

To appreciate the magnitude of the impact, let's make a contrast to a life without sex. Picture yourself as a gene in an asexual lineage, stuck in the humdrum world of cloning. Your mission, handed down by natural selection, is to produce more and faster than your rivals across generations. But here's the kicker: as a tiny gene, you can't go solo. You're one of hundreds or thousands of genes in the genome that must work together to fulfill your evolutionary dream. But the problem is that cloning locks you into the same genetic crew, round after round. In this stagnant genetic neighbourhood, you are chained to collective interest. Your unique flair? Stifled. Your self-interest? Laughed at. You're trapped in an 'all for one, one for all' set-up, where survival means giving your all to the community. Step out of line, and you face the consequences.

But then comes sex. With its magic gene-shuffling powers, sex breaks the shackles and wrecks the cosy little utopia of genetic collectivism. It frees you from the confines of the same old gene community and lets you switch allegiances. Suddenly, a new genetic persona pops up: you're unique, you've got personality, you can pursue your own interests. In this brave new world, individuality isn't just allowed; it's encouraged. Each generation you move to a different genetic neighbourhood, and going solo is more than an option – it's a necessity. In this cut-throat arena of competition, promoting your own interests means embracing your inner Machiavelli. Cooperation, manipulation, cheating, even assassination – everything is on the table when it comes to your survival and reproduction – your evolutionary fitness.

If the above genetic collectivism-individualism trope sounds like human political systems, resist the temptation to compare. Remember, you're a non-moral gene. You have no soul nor free will; you can't make choices against your own interest. Your only concern is your genetic success, as dictated by the demanding steps of evolution: copy yourself, pass it on to your children, outdo your rivals and dominate the gene pool. At times, you will go to astonishing lengths to secure your survival – even resort to murder. Let's dive into one such example.

Male, Female and Beyond

In the laboratory mouse there exists a selfish genetic element known as the t-complex, comprising no fewer than four linked genes. When a heterozygous male (carrying the t-complex in one chromosome but not the other) produces sperm cells, the t-complex manufactures a toxin specifically designed to kill sperm cells that lack the t-complex. As a result, nearly all the sperm end up with the t-complex, defying the expected 50/50 split – according to the classical Mendelian genetics.[18] This is pure self-interest on the part of the gene, even though it reduces the overall number of sperm by half, showing just how spiteful and ruthless genetic competition can be. This is what genes can do, thanks to sex, in breaking up and remixing one gene community with another.

If being a gene hasn't given you warm fuzzies, let's look at it differently – objectively. A clone is like a photocopy of its original, down to the last detail. Every gene in that clone has to play nicely and work together for the greater good of the genome. If one gene decides to slack off, the whole clone is doomed to lose to other, more team-spirited clones. There's no room for a diva gene (like those vicious t-complex genes) – think of it as a race where a selfish gene that ignores the team ends up dragging everyone down and disappears from the gene pool. So while different lines of clones might do battle, genes within a single clone need to get along. No infighting allowed! The only thing genes must do is to cooperate with others, willy-nilly.

This works fantastically in all of us right now. The 37 trillion cells in your body are all clones with the same genetic make-up, coming from one original cell – the zygote.* Except

---

* Today, the symbiotic microbial community in our body is often considered part of us.

for sex cells, all of them work tirelessly at their assigned jobs, despite no chance of passing on their own genes by themselves. Your intestinal cells sift through the food sludge in your gut, your red blood cells play courier with oxygen, your immune cells patrol for invaders, and your neurons and sensory cells collect and process information and make decisions every moment. They all put in the effort without complaint, betting everything on the few sex cells as their proxy to carry their genes forward. If one cell decides to go rogue, your body may face serious trouble such as cancer, which is made of selfish mutant cells that are against the interest of the rest of the body. Here, genetic sameness is nature's superglue, putting a team of twins together for a common goal.

However, sex really stirs up the pot. Sure, it churns out genetic diversity, but it also opens a Pandora's box of conflict.

As we have already discussed, when genes get shuffled into a new genetic neighbourhood every generation, they're no longer stuck in the same old genome, as if they were atomized. They become like rebellious teenagers, each one looking out for number one. Even though genes still need to survive in their gene community and can't just sink or swim on their own, they get some freedom to chase their own dreams – producing as many copies as they can, even at the cost of other genes. In the end, sexual reproduction, as biologist David Haig puts it, produces 'a team of champions' for individual genes, unlike the 'champion teams' emerging from competing clones in asexual reproduction.[19]

Meanwhile, sex encourages genes to evolve a whole arsenal of new strategies for interacting with one another – some constructive, like cooperation, and others ... less so. The darker side – manipulation, cheating, even killing – throws a wrench into the system. This infight among the genes adds an extra layer of cost – call it *the social cost of sex* – by disrupting the peace and harmony of the once-stable gene community.*

• • •

Not all is lost, however – on the contrary, much more is gained. By giving genes individuality, sex breathes life into a stagnant gene community, freeing genes from the chains of an unchanging genome. In their pursuit of self-interest, genes mingle with others in all sorts of ways, which then unleash higher levels of complexity. That's why, for example, sex can drive the formation of new species in multicellular organisms, leading to more complex creatures with all kinds of novel forms and behaviours.[20]

Take body size, for instance. Ever wondered why we have huge mammals like whales, elephants and rhinos, or extinct giants like mammoths, giant beavers and giant ground sloths? It might seem like getting bigger was an evolutionary trend chugging along on autopilot. But recent studies show that large size is actually a by-product of new species emerging.[21] It's really sex that's in the driver's seat behind the evolution of body size. And the story will get even more detailed and delicious when we dive into sexual selection in later chapters.

At this point, you may say, 'Yeah, I get it. Sex drives speciation. But does it work equally for all forms of life?' Great question! The answer, according to Andrew Bush, Gene Hunt and Richard Bambach, is no.

---

* This cost in social cohesion is not part of the cost of meiosis in sexual reproduction. It's the centrepiece of kin selection, an idea we will discuss in later chapters. It also led to Dawkins's conception of the selfish gene.

In a recent study, the trio tallied up marine life and discovered an interesting pattern in biodiversity. They noticed that sedentary species – like corals, sea anemones and clams – live fairly uneventful lives, anchored to rocks or buried in the seafloor. Males of these species release sperm into the water, broadcasting it like shooting in the dark, hoping it finds a target. This sessile existence, even with sex, led to little biodiversity for 450 million years since the Late Ordovician. But when marine life evolved ways to transfer sperm directly from males to females (like in species of arthropods, cephalopods and vertebrates), boom! Biodiversity exploded, tripling the number of genera in just 145 million years since the Cretaceous and Cenozoic – a tenfold increase if we extrapolate their existence to 450 million years.[22]

The moral of the evolutionary story: it's not just about having sex; it's about *how* the sex takes place that really spices things up in the animal kingdom. Just as living like a couch potato won't do wonders for your health, sex without some lively male–female interactions won't do much for biodiversity either. So why do direct interactions between the sexes shake things up so much?

The answer takes us back to the Red Queen hypothesis. Just as it takes two to tango, sex requires the interplay between males and females – often in a negative way – to trigger evolutionary arms races and unleash the power of sex in creating diversity. This is most evident in multicellular organisms,[23] especially in seed plants and animals – species that mostly undergo sexual reproduction – whose sizes range from tiny flowers underfoot to towering redwoods, and from microscopic bugs to colossal blue whales.

To recap, as we know, hostile interactions can drive biodiversity at three levels:

a) Genetic: Individual genes can employ strategies like cheating, manipulation and even killing – and their counter-strategies.
b) Individual: Various interactions between males and females characterized by sexual selection.
c) Species: Interactions between species in ecological communities, including parasitism, predation, competition and more.

These interactions fuel the constant race for survival, driving the incredible diversity we see in nature. Let's zero in on male–female interactions and explore how they generate diversity in many ways and at different levels. This serves as both a highlight and a prelude to sexual selection, a topic we will dive into in later chapters.

If you've ever been interested in ballroom dances like the waltz and foxtrot, you've noticed men taking the lead for better coordination. This norm likely stems from historical patriarchal traditions. But in the evolution of sex, the roles are usually flipped – females take the lead. Why? Because any change in a female's life opens up new reproductive opportunities for males. Parasitic and short-lived, sperm are bound for a quick kamikaze mission: finding and merging with eggs. To succeed, males need to sync up with the females' lifestyles. As the saying goes, beggars can't be choosers, and males have no other option but to go along.

So you should not be surprised to see that even a slight change in a female's behaviour – whether in feeding habit, herding pattern or mate preference – can trigger major shifts in males. Males, meanwhile, are often under pressure to innovate, adapting their bodies and behaviours to boost their chances of pulling off these high-stakes bets. This explains the presence of multiple male types with different mating strategies – in contrast to just one female type – in some species. For example, males of several species of salmon come in three kinds: normal, jack and precocial (or parr). Male side-blotched lizards also have three types: orange, blue and she-male variations. Male ruff sandpipers, a shorebird, diverge into three morphs: grey territorial male, white satellite male and a female-mimicking male called faeder (Figure 3.7). It's a fascinating display of diversity and adaptability, all driven by the quest for reproductive success.

Certainly, females too can gain advantages by adapting their survival and reproductive strategies based on changes in males. When this happens, it sets off an evolutionary *pas de deux*. This back-and-forth dance between males and

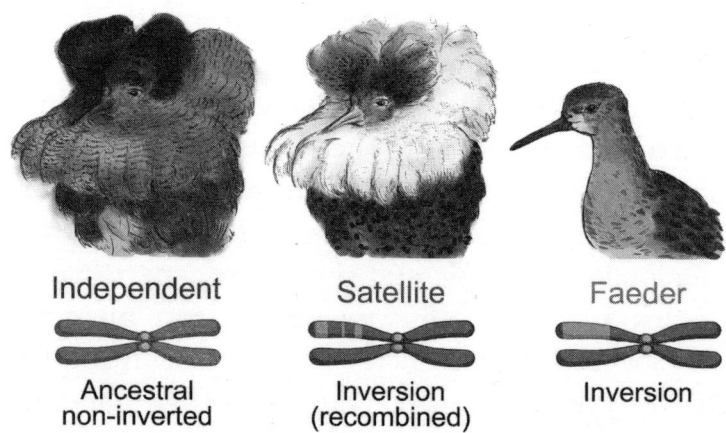

Figure 3.7. Polymorphism in male ruffs: a territorial male (left), a satellite male (middle) and a female-mimicking male called faeder (right) during the mating season. Aside from structural differences in the chromosomes involved, the vast disparities in morphology and social behaviour among the males are linked to the varying expression levels of the *HSD17B2* coding for the enzyme 17-beta hydroxysteroid dehydrogenase 2

females accelerates evolution, resulting in a world brimming with diversity and complexity among sexual species today.

Evolution is like competitive sports, where the goal is to score the highest in Darwinian fitness. But, unlike sports, there are no referees or rulebooks – it's a wild free-for-all. Sex is the ultimate game-changer: it's a powerful tool for innovation, granting life a serious advantage in the survival struggle. By giving an edge in Red Queen-style arms races, sex sparks an explosion of diversity at every level, from individual genes to entire species and beyond. Sex opens the door to a whole new world of possibilities.

Part II
THE UNFOLDING

4

THE MAKINGS AND VARIATIONS OF SEX
Sex Determination and Development

What's the strangest animal in the world? If you guessed the duck-billed platypus, congratulations, you've nailed the top pick according to Google. This consensus tells us two things: first, our animal knowledge tends to spotlight the big and flashy creatures;* second, the platypus is genuinely bizarre. With a snout like a duck, a tail like a beaver and webbed feet like an otter, it's a mash-up that sounds straight out of a mad scientist's lab.

When the first platypus specimen and drawings were sent to the British Museum in 1798, people thought it was a prank. Oxford-educated naturalist George Shaw was convinced it was a hoax, because of its 'perfect resemblance of the beak of a Duck engrafted on a quadruped'. He grabbed a pair of scissors and poked at the snout, searching for stitches. When he couldn't find any, he admitted in awe, 'whatever was possible for Nature to produce has actually been produced'.

The specimen Shaw examined was little more than dry skin. It didn't even have a tongue. So it couldn't offer many clues about the platypus's life. But the more we've gotten to

---

* We often overlook a lot of cool and weird small animals when it comes to this.

know this oddball, the more it's unsettled our sense of normal. For starters, the platypus has a sense of electrolocation, which means it can detect weak electric fields in the water to hunt its prey such as bugs and crayfish. It is venomous, with spurs on the hind feet of the males of the species which, if you are unlucky enough to be on the other end of, cause excruciating pain.

Stranger still, the platypus is a bizarre blend of both bird and mammal. Like a chicken, the female lays two small eggs in the nest, each about the size of a dime, with a leathery shell that protects the yolk until it hatches after ten days. But once the babies hatch they nurse on their mother's milk for about four months, showing off their true mammalian nature.

Strange as all that is, the most recent discovery about the platypus is its true hidden wonder: the surprising character of its sex chromosomes.

• • •

As we know, males and females are biologically defined by the size of the reproductive cells they produce – sperm and eggs – not by characteristics like genitals or hormones. Yet the mechanisms determining and expressing phenotypic sex can be remarkably varied and intricate, often leading to confusion. In this chapter we'll see how sex, as a biological trait, defies simple categorization and frequently challenges our assumptions.

Let's start by examining how genes and chromosomes drive sex determination. In humans, sex is *usually* dictated by sex chromosomes: XX for females and XY for males. These chromosomes house crucial genes that set the foundation for sexual characteristics right from fertilization, laying out the genetic groundwork for the male/female distinction. This

The Makings and Variations of Sex

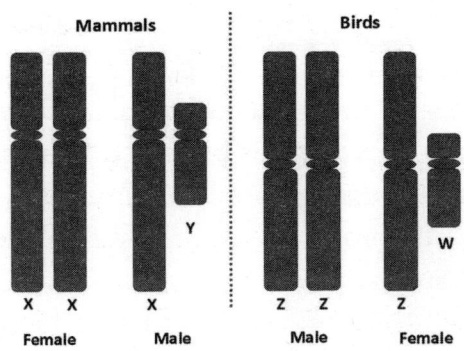

Figure 4.1. Sex chromosomes in mammals and birds

system, called *chromosomal* or *genetic sex determination*, is typical of mammals and birds.

However, the apparent similarity in sex determination between birds and mammals masks a significant difference in their sex chromosomes. In most mammals, including humans, the Y chromosome is smaller than the X chromosome. In birds, the situation is flipped. Male birds have two similar chromosomes, ZZ, while female birds carry two different chromosomes, ZW, with the W being smaller than the Z. Simply put, the ZW system is the inverse of the XY system.

Recalling the two California condors born via virgin birth at San Diego Zoo, both of which turned out to be male, we can now understand this in terms of the sex chromosomes of birds. These birds could only ever have been male, as they were produced by the duplication of genetic material in a haploid egg. Since WW birds, much like YY mammals, cannot survive because they miss some crucial genes on the Z sex chromosome, the only possibility is for the Z to duplicate, resulting in ZZ – male – birds.[1]

Interestingly, some insects, like fruit flies, also rely on the XY system, similar to mammals, rather than the ZW system

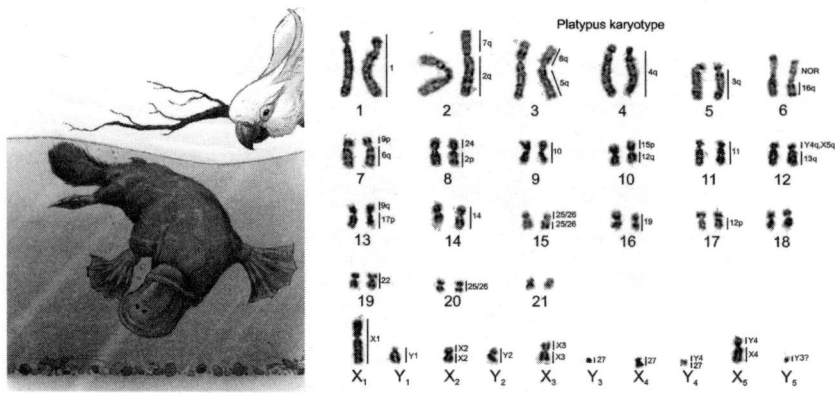

Figure 4.2. Left: Platypus artwork by Nellie Pease. Right: chromosomes in the platypus, noting the five pairs of sex chromosomes at the bottom (Question: is this individual male or female?*)

used by birds. Insects might seem far removed from birds and mammals, which are more closely related to each other in evolution, so how is it that birds and mammals diverge so drastically in sex determination? To dig deeper, we can turn to animals that bridge some characteristics of both birds and mammals – like the platypus.

To wonder at the platypus's peculiar appearance is merely to skim the surface of this intriguing animal. The true marvel is hidden in its cells. The platypus is, by one measure, the sexiest animal around: instead of the typical two sex chromosomes found in most mammals, it boasts a staggering ten. Females have five pairs of X chromosomes, while males have five X and five Y chromosomes. Why they are so 'hyper-sexy' remains a mystery to this day, but how the system works has recently been deciphered by a group of Australian scientists

---

* It's a male because there are five X and five Y chromosomes. A female platypus, on the other hand, has five pairs of X chromosomes.

led by Frank Grützner. Aside from their physical and behavioural traits, the sex chromosomes in the platypus are also like a chimera, with one of them resembling the X in mammals and the other like the Z in birds.[2]

To understand the mechanisms at work in the duck-billed platypus, it's handy to first take a look at species that use radically different sex determination to mammals: social insects such as bees, wasps and ants. In these species, sex is not determined by sex chromosomes but by the number of chromosome sets. Females are diploids with two sets of chromosomes ($2n$), while males are haploids with just one set (n). This system of genetic sex determination is aptly termed *haplo-diploid sex determination*. It represents a unique evolutionary path in the two forms of existence in animals. In this set-up, dominant females (queens), like all other animals with a diploid body, are fully equipped to deliver the haploid egg, wrapping up their life cycle neatly. Haploid males, however, are a different story. Their genomes live part of their lives independently as males and the rest as biological freeloaders, hitching a ride on queens. Lacking the diploid stage makes the male life cycle look like a side loop – a detour – on the main road.

The Western bee, the species that gives us the honey on our breakfast table, offers a glimpse into how sex determination works in a typical haplo-diploid species. Within a bee colony, the queen holds exclusive right over the birth of male or female offspring. This control is exercised by a simple decision: whether to fertilize her eggs with the sperm collected during her mating flights – her flings with males from other colonies.

The queen then lays two types of eggs: unfertilized and fertilized. Unfertilized eggs develop into males. Fertilized eggs, however, undergo a transformation in development, guided by two key genes, *csd* and *fem*. The *csd* gene is nudged to work when it interacts with a different allele (an alternative version of a gene) in its corresponding chromosome. (This interaction is exclusive to females, as male chromosomes are unpaired, existing in single copies.) Once *csd* is activated, it wakes up the dormant *fem* gene. In one fell swoop, *fem* takes charge and flips on the switch of the body's biochemical assembly line, leading to the production of a female.[3]

This elucidates the genetic mechanism through which the queen controls the sex of her offspring, predominantly favouring daughters tasked with managing the myriad household chores in and out of the nest. Despite the colony's flourishing, however, the queen still has to confront yet another formidable obstacle: the looming spectre of rebellion that exists among her daughters. How does such defiance arise?

The key lies in the shake-up that sex brings to gene communities. As we know, sex doesn't just come with the 'cost of meiosis', slashing genetic efficiency in half – it also

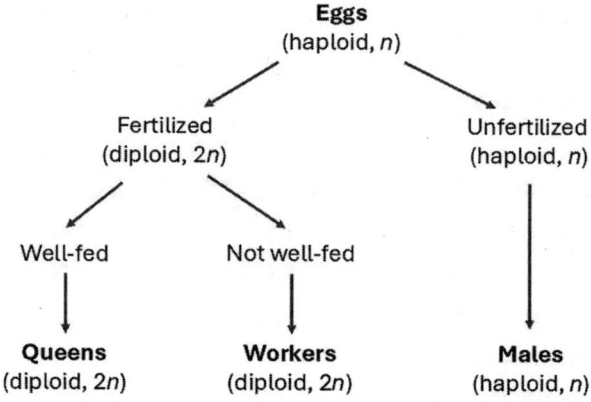

Figure 4.3. Different categories (also known as castes) of bees, wasps and ants, showing how sex is determined

drags in a whole new set of problems. By mixing genes from two parents, it dilutes the genetic similarity between kin, watering down their 'blood' ties. And as that bond weakens, so does the drive to cooperate. Conflict rushes in to fill the gap. So in a hive full of potential rebels – her very own daughters – how does the queen hold on to her throne?

One of the queen's craftiest tricks is chemical castration. She releases a pheromone from a tiny gland near her mouth – the mandibular gland – that works like a charm to stop other females' ovaries from maturing. So even though these daughters have all the right genes to be females, they can't lay eggs, making them, quite literally, unsexed – or intersex, if you prefer – since most of them will never get a shot at reproducing.

But the queen does more than just hold her ground: she tightens her grip by enlisting loyal allies from an unexpected place – her own larvae. These little grubby creatures, though limited in strength, pack a powerful punch with a pheromone called *e*-beta-ocimene, which turbocharges the queen's efforts to prevent her daughters from sexual maturation. Together, their chemical arsenal lets her proclaim, 'The crown shall never be usurped!'[4] Only when Her Majesty meets her end can another female rise from the unsexed ranks, claim the throne and be crowned the new queen.

Fun metaphors and monarchical politics aside, let's get to the point: honeybees show us that the social cost of sex can be a real buzzkill. For the queen bee, her behaviour underscores the headaches of juggling different interests in a community. Can evolution step in and lower the social cost of sex, making the bee colony more peaceful and cooperative?

Absolutely! One strategy is to boost genetic relatedness within the colony to lower the cost of meiosis. It's like transforming half-brothers into full brothers and full brothers into identical twins. This strengthens the social ties that sexual reproduction often strains. It's kin selection in action: the more genes you share (due to inheritance from a common ancestor), the more you're nudged to help each other. After all, it's a lot easier to get along when everyone's in the same genetic team.

How can animals evolve towards promoting their genetic relationships, then? There are two main methods that we know of in nature. Call them 'the ant's way' and 'the naked mole-rat's way', respectively.

As we mentioned earlier, ants, along with their buzzy cousins, bees and wasps, have this neat trick of the haplo-diploid sex determination system, which brings them all closer together genetically. Here's the scoop: a queen mates with a male and starts a colony. She churns out daughters who skip the teenage rebellion phase entirely and dive straight into work. These unsexed daughters take on the roles of workers, nurses or soldiers, hustling non-stop to keep the colony running. Cooperation isn't just their vibe – it's their *raison d'être*. They'll even throw themselves in harm's way to protect their sisters and the queen.

I got an up-close – and painfully personal – demonstration of this loyalty during the summer of 2023. My lawnmower accidently ran over a subterranean yellowjacket nest, and five pint-sized kamikazes swarmed me, stinging my head and back with military precision. The pain lingered for over a week, a not-so-gentle reminder of just how far these little warriors will go for the team.

How does the haplo-diploid sex determination system crank up genetic relatedness? Well, in this set-up, all female nestmates come from the queen. But they're not just regular sisters; they're what I like to call super-sisters. Their genetic closeness is off the charts compared to typical siblings in other animals. Here's how it works: the paternal half of their genetic make-up is identical since it comes from the same male.* On the maternal side, there's an average 50 per cent similarity, thanks to the random shuffle of the queen's chromosomes during egg production. When these two halves are added together, female nestmates share an average of 75 per cent of their genes. That's a huge bonus compared to most other organisms, where full siblings share just 50 per cent of their genes. Thus it's in the daughters' interest to help their sisters in the nest, rather than help their mothers, to whom they relate by only 50 per cent. Now, you can see, the haplo-diploid system is really a clever genetic hack that helps bees, ants and wasps cut down the social cost of sex by boosting genetic relatedness among colony members. That's how kin selection ultimately snuffs out much of the rebellious streak in unsexed daughters.

We call it 'the ant's way' because ants are among the planet's most abundant and successful creatures. With around 30,000 species – half of them still awaiting scientific description – ants dominate in sheer numbers, boasting an estimated 20 quadrillion individuals. Their collective biomass surpasses that of all wild birds and mammals combined,[5] and they play a stunningly wide range of roles in ecosystems as predators, herbivores or scavengers. You can find ants almost everywhere, except in the harshest spots like the Arctic, Antarctic and freezing mountain tops. Given how well ants and other social insects are thriving, it makes sense to

---

* The percentage of shared genes would be lower if the queen mated with several males.

consider that their sex determination system might be a key to their evolutionary success. After all, with a set-up like that, it's no wonder they're winning the survival game.*

The naked mole-rat offers an alternative strategy for managing the genetic disruptions that come with sex. These peculiar-looking rodents live in colonies with a strict division of labour, much like ants, bees and wasps. But instead of relying on the haplo-diploid system, they turn to inbreeding – mating with relatives – to boost genetic relatedness within the colony. This approach lowers the cost of meiosis and reduces the social cost of sex, fostering a level of cooperation that keeps the colony running smoothly.

But inbreeding has its own set of genetic hiccups, as we all know. Using it to slash the social cost of sex is like walking a tightrope while juggling flaming torches. The very fact that naked mole-rats are the only mammal species which has signed up for this crazy ride suggests they might have been trapped in an evolutionary dead end. Apparently, they are losing the war of their own survival while attempting to win the battle against the social cost of sex.

These examples show that no matter how animals crank up their genetic relatedness – whether they go the ant route or the naked mole-rat way – the result is often the same: a tight-knit, intricate society that runs like a well-oiled

---

* Haplodiploidy has been considered by some as an evolutionary dead end, given that it has evolved independently dozens of times without any documented instances of reversal. (Bull, J.J., 1983. *Evolution of Sex Determining Mechanisms* (Menlo Park, CA: Benjamin Cummings.) Recent research, however, shows that it may have advantages in resolving incompatibility in hybrids. (Nouhaud, P., Blanckaert, A., Bank, C. and Kulmuni, J., 2020. Understanding admixture: haplodiploidy to the rescue. *Trends in Ecology & Evolution*, 35(1), pp. 34–42.)

machine. In biology, we have a fancy term for this kind of arrangement: eusociality.* It's when the colony splits up the workload with military precision – some are workers, others defenders, and a lucky few get to focus on reproduction. This division of labour isn't just efficient; it's the secret sauce that turns genetic kinship into a cooperative powerhouse, creating communities so interconnected that they function like a single superorganism.

Humans, in this regard, are the oddballs. Although we too are subject to kin selection, we can also manage to create extensive cooperation beyond our genetic relatives. Instead of relying solely on blood ties, we set up social rules – norms, conventions, customs and laws – to keep everyone in line and working together. But that's a whole different story.

• • •

Although we're used to the idea that genes decide if we're males or females, that's not the only trick nature has up her sleeve in determining this distinction. In many animals, whether you turn out male or female isn't set in stone at the stage of a fertilized egg. Instead, it depends on the environment – things like how long the days are or how high the temperature is. This flexible method is called *environmental sex determination*.

In many reptiles, for instance, the sex of fertilized eggs remains undecided for quite a while. It all depends on the temperature they hatch in, as if nature has a thermostat to control sex the way we use it to turn the heat up and down. There are two main patterns for this kind of sex determination. In Type I, seen in turtles, warmer temperatures make

---

* The prefix 'eu' means 'perfect' or 'good'. So 'eusocial' literally means 'truly or perfectly social'.

more or all females.* Type II, found in some geckos, iguanas and alligators, is a bit different. Here, both high and low temperatures produce females, while the Goldilocks zone – just right in the middle – yields males.[6]

But how – and why – does temperature get the final say on sex? The secret is in the working conditions of the genes involved in sex determination. When the eggs are fertilized, these genes are snoozing. It's only during incubation that they start clocking in for work. But they have a weird quirk: they work differently at different temperatures, almost like they're taking orders from a thermostat.

In the red-eared slider turtle, for example, these genes are like grumpy employees. Keep their 'pay' at a chilly 26°C, and they'll churn out all males. But bump their 'salary' up to a toasty 32°C, and suddenly you've got all females. In between, you get a mix of both. The main supervisor in this operation is a gene called *dmrt1*, acting like a factory boss. On its own, *dmrt1* doesn't do much, but it orders other genes to produce males. To get females, you need to knock *dmrt1* out of commission by targeting its higher-up boss, a gene named *kdm6b*. At lower temperatures, *kdm6b* makes an enzyme that wakes up *dmrt1*, turning it into a male-making machine. But at higher temperatures? Here, *kdm6b* acts rather like how you might if you were guaranteed your boss wasn't going to show up to work. Biologist Chutian Ge and his team have shown that by tinkering with *kdm6b*'s working conditions, they can readily flip over 80 per cent of eggs destined to be males into females.[7]

---

* In tuataras in New Zealand, high temperatures result in more males, and low temperatures make more females.

The Makings and Variations of Sex

Why would animals let the outside world decide the sex of their kids? Though it seems a risky strategy, in reality it gives these species the flexibility to micromanage the male-to-female ratio. Adjusting the proportion of sons and daughters can offer extra fitness perks when the environment is predictable. (Stay tuned for more juicy details about the biological intrigue of sex ratio in Chapter 9.)

But when the environment is unpredictable, messing with the sex ratio can backfire bigtime. Imagine ending up with all males or all females – it would swiftly be game over for sexual reproduction. This gamble for extra benefits can lead to a huge loss of overall fitness. Here again, it's like winning one battle with a highly specific strategical plan – only to end up losing the entire war.

Just like in high-stakes military operations or risky financial investments, trying to predict the unpredictable can be a dicey move. To avoid disaster, you need to secure the basics first with proper numbers of both male and female offspring. That's why, when the environment is a wild card, animals tend to switch to genetic sex determination to play it safe.[8] This evolutionary wisdom is perfectly shown by a copper-coloured Tasmanian lizard called the snow skink.

The snow skink lives from sea level to 3,940-foot-high rocky mountainsides, covering a lot of ground with different temperatures. It has evolved two survival tricks to handle this challenge. First, it keeps and hatches its eggs inside the female's body, rather than leaving them to unfavourable external conditions as in most lizards. But, like all reptiles, the skink is cold-blooded, which means it has to deal with the wild swings of its habitat's temperatures. This brings us to a second strategy.

To handle the shifting conditions, skinks mix up their ways in sex determination. In high-altitude areas, where temperatures can be frigid, erratic and swing like a rollercoaster, they stick to genetic sex determination to keep the sex ratio in check. But down in the lower, warmer regions with more predictable weather, life has its perks – especially for early-born females who get a longer growing season and a hefty lifetime reproduction boost of 50 per cent over their late-born sisters. A savvy skink mom can pull a fast one on nature by hatching more daughters early in the season and saving the sons for later – all with a well-timed sunbathing session. Thus a small tweak can give her a boost to her Darwinian legacy. This two-pronged approach helps the lizard thrive in diverse environments with different temperature challenges.[9]

Today, with global warming in full swing, the clever temperature tricks some species use to decide sex are turning into traps. Unlike the snow skink, most reptiles lay their eggs outside, so even a tiny temperature bump can potentially throw their male-to-female balance off-kilter, making it harder for them to find mates and keep their species going.

Figure 4.4. Snow skink in Tasmania

The Makings and Variations of Sex

If climate change happens faster than they can adapt, we could see some serious population crashes. Now the fate of these creatures is slipping out of their control and straight into our hands.

. . .

Temperature isn't the only weird trick nature uses to decide who gets to be male or female. Take *Bonellia viridis*, a strange little marine worm with an even stranger system – where your sex depends entirely on where you land after birth. These worms are the ultimate couch potatoes, sticking to the rocks where they settle. *Bonellia* larvae start life without a designated sex: they are unsexed and with all possibilities open. If they settle on the ocean floor, they develop into females. But if they encounter a female worm, they're transformed into males, likely spurred by specific pheromones.

Stranger still, most offspring stay inside their mothers' bodies without leaving, and when this happens they also become males, just as they do around any other female, although in this case they're even more noticeably tiny than their external counterparts. Talk about a twist in family roles – these dwarf males produce sperm to fertilize their mom's eggs.[10] And, as if to keep things interesting, some worms stay unsexed, hanging out in sexual limbo, while others develop into intersexes. Still, 17 per cent of them stick with the route familiar to us, relying on genetic sex determination.[11]

Sex determination in *Bonellia* worms is like a wild party where the guest list decides your fate. The star of the show is social context – specifically, pheromones released by females that set the mood. This quirky system, known as social sex determination, lets your surroundings pick your sex, and *Bonellia* worms aren't the only ones playing this game: snails, fishes, insects and even plants are all in on the fun.[12]

This brings back childhood memories for me. My friends used to call me a 'nature freak', partly because I was obsessed with catching swamp eels (*Monopterus albus*) in the rice paddies, ponds and rivers of East China. While prepping these fish for dishes, I noticed that the bigger eels never had eggs, which caused me to wonder if they were all males. As it turns out, my hunch was spot on. Swamp eels start off as females, but around age two or three they kick off their transition and two years later, *voilà* – they're fully fledged males.[13] Of course, I wouldn't understand why until many years later.

Swamp eels aren't the only masters of the sex-change game. Perhaps the reigning champions are clownfish. In the balmy waters of tropical seas, about thirty species of clownfish shimmer with their iconic stripes. These fish don't quite live up to their name. Instead, they have a social order tighter than a military drill team. At the top of the pecking order is the big boss: a large alpha female. Right below her is her loyal sidekick, a slightly smaller but fully grown male. The rest are immature males. This hierarchy keeps the peace with one

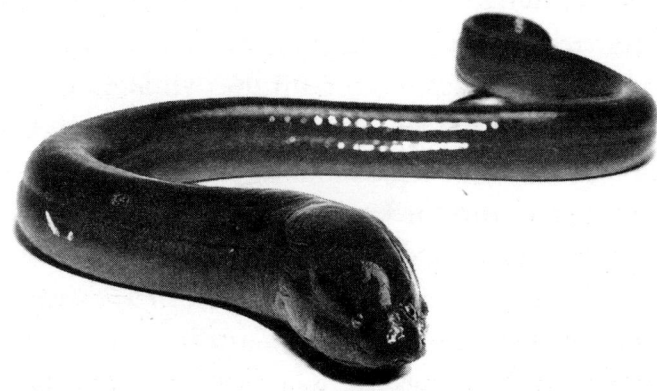

Figure 4.5. A swamp eel

jaw-dropping twist: if the alpha female bites the dust, the top male, the second in rank, begins flipping the script and takes over her role while the top immature male matures.

Over a few months to a year, this new leader undergoes a dramatic transformation. The fish's brain circuits shift gears, sending hormones to its gonads to kick-start the shake-up of its primary sex organs. During this process the testicular tissues are reabsorbed, and the ovaries begin to grow. The male morphs into the dominant female. Once she's settled into her new role, she struts her stuff, bosses the others around, and starts courting the male, also newly minted, for some fishy romance.[14]

Why do clownfish switch sex based on social dynamics? These charismatic reef-dwellers make their home among the stinging tentacles of sea anemones, which offer protection from hungry predators. Wander too far, and you're likely to become someone else's snack. So staying put is better than venturing out for new opportunities. But here's the twist: laying eggs takes a lot of energy, and small fish just can't afford the cost. That's why clownfish start life as immature males – it's a low-investment role that lets them focus on growing. Only when they've climbed the social ladder and reached the top of the social hierarchy do they make the dramatic switch to female. By then they're bigger, better fed, and ready to handle the heavy demands of producing hundreds – sometimes thousands – of eggs. This clever sex switch isn't just a quirk; it's a finely tuned strategy that maximizes reproductive success. In the high-stakes game of evolution, clownfish flip the script to give themselves the best possible shot at leaving behind a legacy.

In contrast to the clownfish world, things run differently in other fishes like gobies, wrasses and swamp eels. In those species, males bring home higher fitness than females. So being male is everybody's ambition. Yet you need size and strength to assume taxing duties like fighting with your rivals and defending your turf, which are a tall order when you're young, small, and frail like a wet noodle. Thus for these species the smarter strategy is to start as a female. Sure, your reproductive potential might be limited at first, but you still gain some fitness and avoid the high-stakes game of defending your nest and children. It's like being a rookie driver on the freeway – sticking to the slow lane is the safer bet. Once you've gained some experience and packed on some muscle, you can merge into the fast lane of reproductive success by swapping to male.[15]

While clownfishes, gobies, wrasses and swamp eels are vastly different species, they share a remarkable trait: the ability to switch sex and enjoy the perks of both worlds. Birds and mammals, however, have to watch from the sidelines, envious of such flexibility. For them, unfortunately, sex is *largely* set in stone by genetics, ultimately. This brings up the question: Do animals with genetic sex determination have any wiggle room to boost their fitness? To answer that we'll need to look under the hood to see what genes are up to when it comes to determining and shaping sex.

. . .

Before the new millennium, we had barely scratched the surface of how genes are involved in the unfolding of sex. But with recent leaps in cell and molecular biology, we've gone from peeking through a keyhole to swinging open the door. Let's take a quick tour and see how genes work their magic to create males and females.

Many animals depend on 'master-switch genes' – genes that work like a switch to turn on the entire blueprint of sex development. In insects like flies, wasps and honeybees, it's the gene *tra* that serves this role.[16] But in most mammals, the master-switch gene is *Sry*.* In other vertebrates such as fish, frogs and birds, the gene *dmrt1* does the heavy lifting.[17] These examples aren't just here to showcase the variety in master-switch genes — they drive home two key points: there are many ways to create males or females, and the evolution of sex determination happens fast. That's why we don't see a one-size-fits-all genetic style for sex determination across all animals.

In contrast, while the master-switch genes that set the whole process in motion can vary, the genes that are directly involved in building male and female gonads are more like ancient relics — they've stayed relatively unchanged over time. These genes, part of what is known as the *doublesex-mab3* (DM) family, are found in both invertebrates and vertebrates,[18] showing just how deeply rooted they are in the evolution of animals.

Perhaps the best way to explain genetic sex determination is by looking at a familiar example: humans. The X chromosome is like a treasure chest of genetic information, packed with around 1,000 working genes. Some of these genes are key players in essential functions, like cognitive ability and sensing red and green colours. On the other hand, the Y chromosome is more of a genetic wasteland, carrying a paltry 70–200 working genes in mammals. In humans, it scrapes by with just seventy-one genes.

---

* For the record, flowering plants also possess genes involved in sex determination. However, unlike animals, no master switch genes have been identified. As for gene-naming conventions, human genes are written in all capital letters, genes in other mammals start with a capital letter followed by lower case, and genes in all other species are written entirely in lower case. Genes with the same letters mean they are the same genes or derived from the same ancestral genes.

On the Origin of Sex

It's fascinating – and a bit tragic – to witness the evolutionary downfall of the Y chromosome. It started out like any other chromosome, lush in viable and vital genes, but over time it's been reduced to what many now see as a genetic junkyard with few working genes. To make matters worse, its sister, the X chromosome, has given it the cold shoulder, refusing to swap genetic material through recombination. This leaves the Y chromosome sidelined and ostracized, slowly deteriorating as Muller's ratchet clicks away. Meanwhile, many of the Y chromosome's hard-working genes have abandoned ship, fleeing to the safety of the X chromosome or other genomic hideouts. This mass exodus has left the Y in men looking a bit like a downsized office after a corporate restructure – shrinking to about a third the size of its once-proud counterpart, the X.

This evolutionary downfall of sex chromosomes has been observed in a wide range of species, including plants, insects, snakes, birds and mammals.[19] In at least ten species of mammals, the Y chromosome has even disappeared completely,[20] raising concerns about the potential loss of the Y chromosome in humans. Should we be worried?

Recent studies throw a bit of sunshine on the Y chromosome's gloomy forecast. In primates, humans included, the gene content on the Y has held steady for a solid 25 million years. Apparently, we have hit the pause button on its genetic decline – for now.* Sure, the Y is still a scrappy little guy, prone

---

* A possible reason for this is genetic recombination within the same chromosome, where segments of DNA are exchanged during meiosis. (Skaletsky, H., Kuroda-Kawaguchi, T., Minx, P.J., Cordum, H.S., Hillier, L., Brown, L.G., Repping, S., Pyntikova, T., Ali, J., Bieri, T. and Chinwalla, A., 2003. The male-specific region of the human Y chromosome is a mosaic of discrete sequence classes. *Nature*, 423(6942), pp. 825–37.)

to mutations and always teetering on the edge of a meltdown. But will it eventually throw in the towel and vanish altogether? The answer, for now, is a cautious shrug: not anytime soon.

Clinging to the Y chromosome like a captain on a sinking ship is the mighty *SRY* gene (or *Sry* in non-human mammals). This master switch gene, a transcription factor by trade,* kick-starts the grand reveal of the male body plan. Think of *SRY* as the project manager of masculinity – barking orders and delegating the tough jobs to its gene crew. Once it's activated, it rallies other key genes to handle the nuts and bolts of male development. But without *SRY* calling the shots, the system sticks to its original blueprint: building a female body by default.

The presence or absence of *SRY* is informative but lacks detail, akin to seeing just the title of a book or a movie. What captures our interest lies in the content – the story with all its intricacies, uncertainties, unexpected twists and, most important of all, takeaways. This is true for the genes involved in the making of sex as well. The real marvel is in how the manufacturing process unfolds, revealing the elements that make it captivating. This process, known as sex development, is incredibly complex, involving no fewer than seventy-five genes in many regular chromosomes[21] – technically known as autosomes – in addition to the sex chromosomes in humans.† Unfortunately, delving into the specific

---

* A transcription factor is a special type of gene that controls how other, protein-coding genes work.

† The term 'sex chromosome' was coined in the early twentieth century, when these chromosomes were first discovered to be associated with male or female. However, recent discoveries have revealed that many genes involved in sex determination and development are actually located on autosomes. This has led researchers like Sarah Richardson to argue that the designation 'sex chromosome' may be a misnomer. (Richardson, S., 2011. *Sex Itself* (Chicago: University of Chicago Press.)

details would require a whole book. Here, we can only provide a minimalist approach to touch on what happens to a few key genes in humans, typical of mammals.*

The journey of a fertilized human egg begins with a frenzy of rapid cell divisions: one cell splits into two, two into four, and things quickly snowball from there. For about five weeks, though, all these cells look like identical clones, and the embryo itself is a sexless blank slate with the potential (or rather, the bipotential) to develop into either male or female, regardless of what the sex chromosomes have to say.

Sex determination kicks off around weeks six to seven of embryonic development. If there's a Y chromosome in the mix, the *SRY* gene steps in like the project manager. It flips the switch on a key assistant, the *SOX9* gene, which then brings in another player, *DMRT1*, to get the job done. Together, this gene squad activates the biological machinery for male development. What was once a blank slate of undifferentiated cells now begins its transformation into testicular tissue – the first step in building a male reproductive system.

Not all mammals use the *Sry* gene as the master switch, though. The platypus, for example, lacks an *Sry* gene altogether. Instead, this peculiar creature relies on the *Amh* gene, which regulates the production of anti-Müllerian hormone, to guide sex development towards male.[22] The *Dmrt1* gene might also play a crucial role in determining sex in the platypus, adding another layer of complexity to its genetic toolkit.

In the human embryo with XX chromosomes, the absence of *SRY* allows the gene *WNT4* to take the reins and

---

* You can get a detailed account about what happens in genes, cells, organs and body structures from Richard Prum's 2023 book, *Performance All the Way Down* (Chicago: University of Chicago Press).

declare: 'Let There Be a Female!' Under the influence of *WNT4* (in conjunction with another gene, *RSPO1*), the gene *FOXL2* gets turned on. This then sets in motion the biological machinery to build the ovaries from clusters of undifferentiated cells. By the sixth week of gestation, the differences between male and female embryos, which are about the size of a pea, become visible even to the naked eye.[23]

From weeks eight to twelve and beyond, the embryo dives into the third and most spectacular stage: sex differentiation. This is when dozens of additional genes fire up one after another, setting off a chain reaction that ripples through the body, shaping a whole range of tissues and organs linked to the gonads. In the end, this intricate genetic symphony produces a fully formed male or female baby.

• • •

As we've come to see, sex determination and development in animals is anything but simple. It's a genetic juggling act involving many dozens of genes (at the very least) that add a great deal of flexibility, allowing animals to adapt to all sorts of environmental twists and turns, persisting regardless of what nature throws at them. In this light, organisms reveal their true purpose argued for by Richard Dawkins: they're vehicles for passing on genes. Sex, then, isn't the end goal – it's a versatile tool for life to explore and adapt to endless possibilities. That's why every traditional view of biological sex – whether it's based on chromosomes, genes, hormones, brain wirings or internal and external anatomies – faces a slew of exceptions and challenges. Even the most basic definition of sex – sperm- versus egg-producer – features more variations and exceptions than we once thought. The takeaway here? Sex is fluid.

In the last chapter we dug into how sex disrupts the gene community, adding a social cost on top of the classic 'twofold cost of sex' or the 'cost of meiosis'. But, as the Chinese wisely say, 'Crises also bring opportunities.' The social cost of sex cracks opens a fascinating evolutionary loophole, paving the way for a diverse presentation of sexual traits – including intersex – to flourish and adapt through the powerhouse of evolution: *kin selection*. It lets you pass on your genes not just by having offspring yourself but also by lending a helping hand to your relatives, who share many of your genes.

Think of it like competing in a mega-event – say, a triathlon, heptathlon or decathlon. Winning gold doesn't necessarily mean crushing every single event; it's about scoring enough points across the board to secure the top spot. Kin selection works the same way. You don't have to reproduce directly to win the evolutionary game. If your genes succeed in your relatives, you're still in the running for the ultimate prize: genetic immortality.

Just look at a beehive, an ant nest or a naked mole-rat colony. Most members are unsexed workers and warriors. Disregarding the potential, they are basically intersex individuals who don't waste time with reproduction – they leave that to the queen. And her offspring don't just carry her genes; they carry the genes of her loyal, related colony members. In a way, it's just like how the parts of your body work in unison: liver, heart, lungs, spleen, guts and brain don't reproduce on their own – they rely on the gonads, all thanks to kin selection. For these eusocial creatures, the entire colony becomes a single, well-oiled machine – a living, breathing and maybe just a little spooky version of Dawkins's vehicle for genes.

Though not always as flashy, kin selection shows its power among birds and mammals too. Take the Florida scrub jay, one of about 300 bird species where young birds may act as 'helpers-at-the-nest', choosing to assist their parents in raising younger siblings rather than immediately reproducing themselves. Then there's the Belding's ground squirrel, which risks its life to sound alarm calls and warn family members of approaching predators. Monkeys and apes often ally and share precious food with close relatives, and let's not forget the naked mole-rat.

These are just a few classic examples of animals giving up their own benefits – from minor sacrifices to major losses including reproduction and life – to boost the family gene pool. It's the ultimate 'team player' move, nature-style. And we humans aren't all that different. Every time we fork over cash to help a sibling or scribble our kids' names into our will, we're playing the same game. Kin selection is biology's version of splitting the dinner bill – painful in the moment, but somehow satisfying knowing it's all for the family – or more precisely, shared genes.

This knowledge about sexual reproduction challenges the concept of 'biological sex'. With kin selection in the mix, biological sex isn't just a matter of splitting the world into sperm-producers and egg-makers; it opens up room for a whole spectrum of intersex conditions as a valid evolutionary strategy for passing down your genes, with or without producing viable sperm or eggs and with or without direct reproduction.*

If we still try to define biological sex just by the size of gametes, we're missing out on the full spectrum of clever

---

* For individuals that promote fitness by aiding genetic relatives, their existence hinges on the reproductive success of others. Thus we can interpret intersex individuals with reduced fertility as an adaptive strategy coexisting alongside traditional males and females in biological terms.

adaptive strategies that sexual reproduction has cooked up. It's time for an update – one that includes intersex identities and a wide range of sexuality as natural, adaptive parts of life's diversity. In this light, the old notion of a strict male–female binary starts to feel not just limited, but downright outdated.

. . .

This brings us to the topic of intersex in humans.* Though a small minority, intersex individuals persist in all cultures and have been recognized and documented throughout history in both Eastern and Western societies. In ancient China, for example, they often piqued public curiosity and were sometimes recruited as court entertainers. In Greek mythology, Hermaphroditus, the intersex 'child' of Hermes and Aphrodite, was a popular figure. This theme was frequently depicted in Greek and Roman art, often showing a blend of male and female features. Notably, classical writers like Plato, Aristotle, Ovid and Pliny also discussed intersex animals and humans.[24]

Despite the long historical records, the adaptive value of human sex variation is not yet well understood, let alone widely appreciated or embraced. Even today, deviations from normative male and female types are still pooled together and labelled under the clinical term 'disorders of sex development' (DSD).

One common male DSD is hypospadias, a condition indicated by the opening of the penis away from the tip.

---

* It's a popular confusion to mix 'intersex' with 'hermaphroditism', because they are fundamentally different in biology. Hermaphroditism is a species-wide adaptive strategy, whereas intersex mostly refers to some individuals displaying both male and female reproductive parts in a species primarily composed of males and females.

The Makings and Variations of Sex

This affects around one in 250 to 350 male infants, or up to 15.7 million men of all ages worldwide as of 2022, to highlight its prevalence. More prominent DSDs involve cases of 'truly ambiguous external genitalia', which occur in about one in 4,500 infants globally.[25] Incidence rates vary by region, with Saudi Arabia reporting one in 2,500 births and Egypt one in 3,000.[26]

Probably the most common DSD condition is known as late-onset congenital adrenal hyperplasia (CAH). It affects the adrenal glands, which produce hormones like cortisol and androgens (male sex hormones). Unlike the classic form, which appears at birth, late-onset CAH usually shows up during adolescence or adulthood. It's caused by a partial deficiency of an enzyme called 21-hydroxylase, which leads to an overproduction of androgens. Symptoms vary but often include irregular menstrual cycles, excess body hair, early beard growth and sometimes infertility.

Better-known cases of DSDs involve unusual sex chromosome combinations, like X0 (Turner syndrome, with only one sex chromosome), XXY (Klinefelter syndrome), XXX (trisomy), XYY, and even rarer combinations like XXXY or XXXXY.* While individuals with these variations often face infertility, some are fully fertile or nearly so (such as XXX and XYY). And, as we already know, even those who are completely infertile can still pass on their genes through helping their genetic relatives. These persistent variations in human sex chromosomes don't just challenge the tidy binary view of sex – they also drop tantalizing clues about their potential evolutionary advantages, mysteries science has yet to fully unravel.

---

* These rare sex chromosome patterns (or karyotypes) are also present in other mammals.

While many cases of DSDs are tied to genetic mutations – ranging from small tweaks to major overhauls – variation in sex characteristics isn't always about mutations. Even without genetic glitches, the many dozens of genes involved in sex development in mammals can work in an enormous number of combinations, creating a wide range of natural variation in both males and females including intersex conditions – such as those with mismatch between chromosome and anatomical sex.[27] (Such mismatch accounts for 3–6 per cent of birds according to a 2025 study in Australia.[28])

Table 4.1 offers a rough snapshot of intersex conditions in humans. Rates can vary depending on how they're diagnosed. But even with a cautious estimate of 1 per cent[29] – instead of the 1.728 per cent listed in the table – that still amounts to nearly 700,000 in the UK, 3.4 million in the US, or more than 80 million people in the world of 8 billion people. It's a reminder of just how common intersex individuals are, even if their stories mostly go unheard.

Causes	per cent
Non-normative sex chromosomes	0.193
Androgen insensitivity syndromes (AIS)	0.0076
Classic CAH (congenital adrenal hyperplasia)	0.00836
Late onset CAH	1.5
Vaginal agenesis	0.0169
True hermaphrodites	0.0012
Idiopathic	0.0009
Total	1.728

Table 4.1. DSD/intersex conditions and their frequencies in humans (modified from A. Fausto-Sterling, 2020[30])

Because we're still unravelling the evolutionary significance of intersex, slapping the label of 'disorders of sexual development' on these variations is not just premature – it's flat out wrong. Tragically, many people with these conditions – especially young children – are subjected to surgeries simply for falling outside the anatomical norms of sex organs.

Evolutionary biologist Richard Prum puts it bluntly: '[A] consistent number of individuals simply defy binary classification,' and most of these variations are neither harmful nor debilitating. Yet they're often treated as medical necessities. Prum and many feminist scientists are calling for a shift in language, advocating that 'disorders of sex development' be replaced with 'differences in sexual development'.[31] The new term not only avoids the undeserved stigma but is far more scientifically accurate.

Intersex conditions, though relatively common, often fly under the radar – until the spotlight shifts to sex identity, as it often does in high-stakes sports. Enter Caster Semenya, a South African middle-distance runner who became a household name in Britain and many other countries after winning the women's 800-metre event at the 2009 Berlin World Athletic Championships. At just eighteen years old, Semenya found herself at the centre of a global controversy about her sex identity. The uproar wasn't subtle. With her muscular build, strong jawline, straight waist and flat chest, Semenya's appearance fuelled doubts and provoked outrage. One competitor fumed publicly, 'For me, she's not a woman. She is a man.'[32]

Semenya's eligibility quickly became the lightning rod for debate, prompting the International Association of Athletics Federations (IAAF) to act with uncharacteristic speed. A mere three hours after her 800-metre triumph, they announced an investigation into her case. The results? Tests revealed an unusually high testosterone level in her bloodstream, leading to an eleven-month ban, ostensibly for doping.

But was Semenya really doping? The IAAF played coy, refusing to disclose the specifics. However, leaked information painted a different picture: the controversy wasn't about doping but the ambiguity of her sex identity. This sparked deeper, thornier questions. Was Semenya male or female? Should she compete in women's events? The answers, it turned out, were anything but simple.

Fair play is the bedrock of sports, and it's the rationale behind separating men's and women's competitions. But before the 1960s, the International Olympic Committee (IOC) enforced this principle with a jaw-droppingly invasive method: the infamous 'nude parades'. Female athletes were required to walk unclothed before a panel of medical judges, who would then give their verdict on the athletes' sex. Unsurprisingly, this practice sparked outrage – not just for its blatant violation of privacy but also for its unreliability and susceptibility to cheating.* Faced with growing ethical concerns and a public backlash, the IOC eventually abandoned nude parades in favour of a more 'scientific' approach: testing athletes' sex chromosomes.

The chromosome test seems straightforward. In a typical woman's body cell, only one of the two X chromosomes is

---

* For the Soviet Union and Eastern European nations at the time, sports were a matter of national pride and a powerful propaganda tool. Winning medals wasn't just about athletic achievement; it was a showcase of socialism's supposed superiority over Western capitalism. Athletes were often encouraged or even coerced to do whatever it took to bring home victories for the 'honour' of their country.

The Makings and Variations of Sex

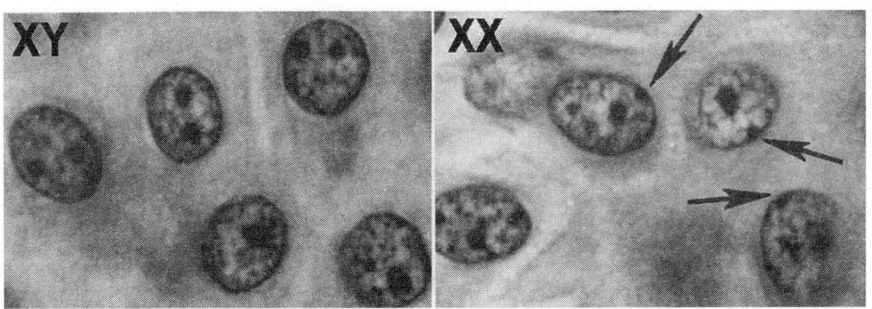

Figure 4.6. Barr bodies used in checking for the female sex

active, while the other is shut down and shrinks to a fraction of its original size. This condensed X chromosome, known as a Barr body, was discovered by Canadian biologist Murray Barr and his graduate student Ewart George Bertram in 1949. The presence of a Barr body is a key indicator that a cell has two X chromosomes (Figure 4.6).

But here's the catch: the Barr body test, initially hailed as a breakthrough, had a glaring flaw. Designed to detect an inactivated X chromosome (the hallmark of females with two X chromosomes), the test could confirm extra X chromosomes but often missed a critical detail – the presence of a Y chromosome. This loophole left the test unable to consistently determine an athlete's functional sex identity.

Would testing women athletes for the *SRY* gene – the gene that triggers male development and boosts testosterone levels – yield more accurate results? The IOC thought so at first, but this approach quickly hit a snag. During the 1996 Atlanta Olympics, for example, eight anatomically female athletes were found to carry the *SRY* gene. The reason? They were naturally intersex, with XY chromosomes. Some of them had a condition called androgen insensitivity syndrome (AIS). Despite having a typically male genetic make-up (XY), individuals with AIS are unresponsive to androgens – the hormones responsible for developing male physical traits. As a result, their bodies follow a female developmental pathway, including forming female-typical external genitalia. This biological twist set off a firestorm of debate: should women with AIS be allowed to compete in women's sports, or did their genetics give them an unfair edge?

When none of the traditional methods proved fully reliable, sports authorities like the IOC and IAAF shifted their focus to serum testosterone levels, which could also help detect the use of performance-enhancing drugs. In women's sports, athletes are now required to keep their testosterone levels below the minimum typically seen in men to qualify for competition.[33] It was this protocol that led to the suspension of Caster Semenya after her testosterone levels were found to exceed the set threshold.

Testosterone levels, however, are a shaky foundation for drawing clear lines in the sand. They can't definitively reveal whether a woman is doping, has an *SRY* gene, or simply falls within the natural hormonal overlap between sexes. Like height or weight, testosterone levels vary widely and may blur the boundary between male and female. Setting a hard cut-off point is a statistical minefield, bound to produce false positives and false negatives.*

Eventually, Semenya was identified as intersex. Though outwardly appearing female, she has XY chromosomes and internal testes that never descended.[34] This was why she had naturally elevated testosterone levels. A similar controversy unfolded at the 2024 Paris Olympics involving Algerian boxer Imane Khelif, who faced intense scrutiny in the welterweight

---

* Also known as Type I and Type II errors in statistics, respectively.

division – not due to any confirmed issue with her sex identity, but because some observers suspected she might have XY chromosomes and perceived her appearance as more typically male.

What these cases show us is that traits like body features, X chromosomes, the *SRY* gene and testosterone levels are all connected to sex, but none of them alone can truly define it. This brings us back to our central issue here: the concept of biological sex. If we continue to rely on the traditional definition of male and female based solely on sperm or egg production, we risk excluding those who don't typically produce either – many intersex individuals.

It's hard not to appreciate the good intentions and great efforts of sports authorities to ensure fair play in women's sports. Yet their attempts have faltered because they are trying to accomplish an impossible task: dividing people into binary categories when human sex identity is more complex.

5

THE GENDER BENDER

Gender and Adaptation

Though early in the morning, the father is already on edge. Just yesterday, one of his four children was taken from him, and the loss has hollowed him out. Despite getting little sleep, his senses are still razor-sharp, scanning the world around him for the slightest hint of danger. His remaining three kids are nearby, snacking and tumbling through a landscape filled with lilies, hyacinths and water lettuces. But this picturesque scene seems too beautiful to be trusted: the father isn't fooled. Danger has a way of hiding in plain sight.

At nine o'clock, it appears.

A figure, female and unfamiliar, lingers across the shore. The father freezes. Something about her – too still, too focused – sets off alarms in his head. His heart races. Without a second's hesitation, he calls to his children, herding them together, urging them to move. But they aren't fast enough.

And she is approaching.

Her movements are deliberate, relentless. The father puffs himself up, shouts, makes every attempt to frighten her away. She doesn't flinch, doesn't slow. Instead she circles, sharp and calculating, until she finds her moment. With ruthless precision, she breaks through his defences, cutting him off from his children.

The Gender Bender

Panic erupts. The kids scatter, little bodies darting for cover, but the stranger is quicker. One by one, she hunts them down. The father lunges, desperate to intervene, but he is no match for her cunning. The first one falls, then another, and finally the last. Gone.

The father's world shatters. He stands there, wrecked, the scent of death thick around him. His grief boils into fury as the stranger turns her attention to him. But this time there is no violence in her approach. She hesitates as she lays her eyes on a new conquest, her posture softer – an overture for a love affair.

He resists, but despair has worn him down. Against his better judgement, he doesn't flee. A strange partnership emerges between the two. Soon, new children are born. With them, the father begins to hope again.

Then, just as suddenly as she arrived, the female departs. No warning. No farewell. The father is alone once more, left with the daunting task of raising his and her offspring, haunted by the memory of the ones he has lost.

This harrowing tale is the love life of the jacana – a water bird that flips the dating rulebook on its head every breeding season. Admittedly, I've sprinkled in a bit of flair for the drama, but in essence this tale is straight out of a naturalist's notebook.

Jacanas are funky-looking birds that hang out in warm spots across Africa, the Americas, Australia and Eurasia. One of the species – the pheasant-tailed jacana – stole my heart during a college field trip to Taihu Lake in East China and turned me into a lifelong bird nerd. Although jacanas have an almost mystical way of evoking serenity for me, what truly captured me was how these birds turn gender roles completely on their head.

Out of the eight existing jacana species, seven have reversed their gender roles.* In the jacana world, the males are the ones who build nests, sit on eggs and play Mr Mom to the chicks. These dads are pros – they keep a close watch on the little ones, using their wings like umbrellas to shield them from rain and cold. If danger is on the horizon, they don't mess around. They'll scoop up the chicks and carry them to safety, just like we humans would grab our kids and run.

This is a world where the ladies call the shots. Forget about males puffing up their feathers – here the females run the show with unapologetic swagger. In some species, female jacanas are a whopping 50 per cent bigger than the guys and guard their turf (and the males within it) like seasoned warriors, wielding blade-shaped wing tips as if they were auditioning for *Gladiator*.

When it comes to love and war, they don't show mercy in their conquest either. If a female jacana spots an attractive male busy playing nanny to someone else's kids, she simply swoops in, channels her inner Medea and mercilessly eliminates the rival's chicks. Cold? Absolutely. Effective? You bet. With the nest cleared of all evidence of her competition, she mates with the male, lays her eggs and then sets off to scout for her next trophy husband.

The wild drama within these conquests plays out like a soap opera, with twists and turns that would make any TV producer envious. In some jacana species, the females might hang around and pitch in a little with parenting, but it's nothing compared to the heavy lifting the males do.[1]

---

* Although biologists commonly use the term 'sex-role reversal' to describe this phenomenon, this book will instead use 'gender reversal' to emphasize the distinction between sex and gender. The rationale for this choice will become evident later in this chapter.

The Gender Bender

Figure 5.1. Left: a female northern jacana with wing spurs. Right: a male pheasant-tailed jacana carrying his chicks

Gender role swaps from the male/female stereotypes we're familiar with aren't rare in nature. In birds alone, jacanas are among the thirty-five species – mostly wading birds like phalaropes, plovers and sandpipers – where males and females flip the script. And if we count species with only partial role reversal, the list gets much longer. In emus, ostriches and penguins, for example, males take on a big role in hatching, caring for and protecting chicks.

In the fish world, gender-bending behaviours aren't rare either. For instance, the Australian seahorse typically forms monogamous pairs, just like a lot of human couples. And every day, the female visits the same male for a few minutes of 'morning greetings' – a special courtship ritual where they swim together, showing off their vibrant colours. If the male is receptive to her mating move, the female transfers her eggs into his brood pouch, where he fertilizes and nurtures them.

Intriguingly, this morning ritual continues until the male's pregnancy is over.[2]

These curious examples lead us to the central question of this chapter: why do animals swap gender roles? To tackle that, we first need to understand what gender means in biological terms – and whether it offers any adaptive advantage. But before we dive into the science, let's take a step back. What exactly does gender mean in today's popular understanding?

Gender, according to the United Nations, 'refers to the social attributes and opportunities associated with being male and female and the relationships between women and men and girls and boys, as well as the relations between women and those between men'.[3] Unfortunately, this definition comes with many problems. It's too narrow, stuck on the idea of men and women, boys and girls, while ignoring everyone who falls in between on the gender spectrum. It's exclusive, laser-focused on behaviour and social relationship but blind to other key players like morphology, physiology, genetics, hormones and neurons. It's also hopelessly homocentric, cut off from our evolutionary connections with other species. In short, it fails to break away from old-school gender stereotypes.*

To do this, we first need to understand what gender really means across the broad sweep of the biological world – not just within narrow human definitions. But here's the kicker: even

---

* The UN's definition of gender reflects a dated nature–nurture dichotomy, believing that culture and biology are separate. This view has been challenged since the 1990s, particularly in animal behaviour, evolutionary biology and epigenetics. Some good reviews of the history and philosophy of gender can be found in *The New Gender Paradox: Fragmentation and Persistence of the Binary* (Cambridge: Polity Press, 2022) by sociologist Judith Lorber and *Gender Diversity: Crosscultural Variations* (Long Grove, IL: Waveland Press, 2014) by anthropologist Serena Nanda.

though many animals routinely flip the script on gender roles, scientists still haven't landed on a clear, objective definition of gender.* Today, despite much progress made, gender remains a controversial concept both in academia and society.[4]

Sex and gender are related but distinct. Biological sex refers to an organism's reproductive role – male if it produces sperm, female if it produces eggs. (For simplicity, we'll set aside the intersex individuals who may not directly participate in reproduction.) Gender, as we've seen in jacanas, often goes beyond biological sex, encompassing behaviours, roles and social dynamics. The difference becomes clearer when we compare sex reversal with gender role reversal in the animal kingdom. Some species – like clownfish, gobies and swamp eels – can actually switch their biological sex during their lives. Others – like jacanas, seahorses and pipefish – don't change sex but reverse gender roles: for instance, males provide care, while females compete for mates.

These examples drive home a key point: gender and sex are not the same thing. But that raises the next big question – what is gender, really, and why do some animals bend its rules, flipping roles or breaking the mould altogether? In popular discourse, gender is often treated as a social construct. What's missing, though, is a more basic understanding of gender as a *biological* construct – something evolutionary biology can help uncover. That's where I come in. As such, we'll stay grounded in biology and leave the deeper and richer philosophical and cultural aspects to the humanities and social sciences.

● ● ●

---

* A major reason is the lack of objectivity, which renders the UN's definition of gender difficult to gain broad acceptance for the scientific community.

We'll begin our journey in Brazil. In many dry caves there, you'll encounter a flea-like insect called *Neotrogla*, often found banqueting on bat carcasses and faeces (also called guano). This is no ordinary flea though: in the *Neotrogla*, egg-producing females have penises and sperm-producing males have vaginas, the exact opposite of what you might expect. Moreover, during mating females take charge by mounting on males and inserting their genitals to collect sperm. Amazingly, lovemaking in this tiny insect can last a mind-boggling forty to seventy hours.[5]

Why is there such a radical change in gender, including a swap of external sex organs, in this species? The answer lies in their ecological conditions. Dry caves lack resources, with little that can be considered food – except for occasional bat carcasses that *Neotrogla* can feast on. While bat guano offers some consistency in food supply, it's low in both calories and protein. Females of this species face a greater challenge than males because producing eggs demands a large amount of nutrition. To overcome this challenge, females turn to males to fill the nutritional deficit. But how?

In the pursuit of attracting females, male *Neotrogla* produce spermatophores, a term for sperm mixed in a protein-rich seminal fluid used as a nutritional incentive to attract the opposite sex. Females respond to the perk in three ways, all through copulation. First, they use their penises to siphon nutrients from males. Females also engage in fierce competition among themselves for more protein by mating with multiple males. This intense competition often leaves males with little time between copulations to replenish their spermatophores. Finally, females mate with males for an extended period of time so that they can mop up every morsel of seminal fluid inside the male's body. Males, in the meantime, can't withdraw without surrendering everything to their assertive mates.[6] With females this relentless at extracting resources, you might wonder if the males can even tell the difference between mating and being mugged. The good news? Unlike some unlucky, cannibalized spiders, *Neotrogla* males walk away from the deed alive – every time. You may be thinking, this is just one insect, and a specific set of circumstances that caused the species to have reversed sex organs. Not so!

If the cave-dwelling *Neotrogla* seems a bit too obscure an animal, consider this better-known example: the spotted hyena of Sub-Saharan Africa. Hyenas live in matriarchal societies – clans that usually have around twenty-nine members but can grow as big as 130 – where the ladies call the shots. The guys? They're the ones who pack up and move out when they hit adulthood. Adult females are roughly 10 per cent brawnier than the males, making it clear who's in charge. Female hyenas keep a strict pecking order among themselves, but when it comes to males, there's no competition – females almost always outrank them. They get first pick at mealtime, no questions asked. As for the newest male to join the clan, he is automatically shuffled to the bottom of the social ladder.

Many people are familiar with the flipped gender roles in hyena societies, but it goes further than that. In a truly fascinating twist, female spotted hyenas sport an external sex organ called a pseudo-penis,* which looks strikingly similar to the real deal in males. To complete the look, they even have a scrotal sac, though it is a bit smaller than the males'. The resemblance is so uncanny that telling male and female

---

* It's really the clitoris and some of the labial tissue that form the structure.

hyenas apart can be tricky. But these unique gender features come with a hefty price for the females. They lose their vaginal opening, which can make mating and birthing difficult. In fact, it can lead to complications during delivery, including an elevated risk of cub suffocation.[7] Given the problems, why would evolution push females to adopt these masculine features in the first place?

The reason behind this gender-bending adaptation is similar to the challenge faced by *Neotrogla*: fierce competition for food. Spotted hyenas often live in large clans, but they don't hunt as one big mob. Instead they split into smaller squads,* a clever strategy to cover more ground and boost their odds of a successful kill.

These hunting packs are often so small that they consist of just one or two individuals. But when a kill is made, up to fifty-six clan members can crash the feast. Even a large antelope carcass can be picked clean in just thirty minutes. It's like preparing a BBQ for ten people, only to have 100 show up. You can imagine the scramble for a bite!

To meet the intense nutritional demands of pregnancy in this cut-throat environment, female hyenas have evolved into larger, more aggressive powerhouses, outmuscling the males – especially when it's mealtime. This strategy eliminates competition, as even the lowest-ranking female outranks the highest-ranking male. To bulk up, females crank out more IGF-1, a growth hormone that is also found in the milk we drink. On top of that, dominant females experience a surge in androgen levels during late pregnancy, which enhances their competitive edge and aggressive behaviour.[8] But this

---

* It is formally known as a fission-fusion society, common in primates but rare in other mammals.

The Gender Bender

Figure 5.2. Males, females and even cubs in the spotted hyena have conspicuous external genitalia

hormonal boost comes with a cost: elevated prenatal androgens can masculinize female offspring, contributing to the development of male-like external genitalia,* a structure that makes labour exceptionally difficult and risky.

Although the cave insect *Neotrogla* and the spotted hyena couldn't be more different, their shared evolutionary destiny of gender reversal sends two clear messages. First, that *biological sex is not determined by external sex organs*. And second, *gender isn't just about behaviour; it also involves corresponding changes in morphology, physiology and life history*.

• • •

---

* This issue remains unresolved. Some researchers believe that the presence of male-like external genitalia in female hyenas may function as a mechanism for signalling dominance, deceiving rivals or facilitating mate choice. Note that these hypotheses are not necessarily mutually exclusive.

Even so, the biology of gender goes deeper. For this, let's take a trip to Florida. Most of the state sits less than 12 feet above sea level. This flat, open landscape is great for farming but also a magnet for pollution run-off, especially pesticides that leach into the soil during farming.

In the 1990s a team led by biologist Louis Guillette Jr stumbled upon a jaw-dropping discovery in some Florida lakes: young male alligators were developing female-like traits, including what the scientists delicately described as 'abnormally small phalli' (which, in plain terms, means tiny penises). The prime suspects? Dicofol and DDT – pesticides designed to mess with insect reproduction but apparently just as good at scrambling alligator hormones. Both chemicals mimic oestrogen in animals, wreaking havoc on the natural balance of sex development.

Upon closer inspection, the researchers found that male alligators in the most polluted lakes had oestrogen levels up to three times higher than those in clean lakes, and their testosterone levels were markedly lower. Even more troubling, this hormonal imbalance seemed irreversible, making 'normal sexual maturation' improbable.[9] And just so we're clear, both males and females have testosterone and oestrogen; it's the ratio, not the type, that makes the difference. In any case, this alligator news hit home for some Florida locals, who couldn't help but worry that their polluted drinking water might be messing with their own manliness.*

---

* Pesticides and fungicides are loaded with environmental pollutants like polychlorinated biphenyls (PCBs), bisphenol-A (BPA) and dioxins. Studies using animal models show these chemicals can mess with sexual development by latching on to steroid receptors. This could be a reason why cases of hypospadias – the intersex condition we talked about in the last chapter – have doubled in the US between 1970 and 1997. (Paulozzi, L.J., Erickson, J.D. and Jackson, R.J., 1997. Hypospadias trends in two US surveillance systems. *Pediatrics*, 100(5), pp. 831–4.)

The solution to this crisis of manhood seems obvious in theory: just get rid of the pollutants. But the deeper, more scientific question of how hormones influence gender's biological aspects is a tougher nut to crack. Though this biology puzzle was posed as early as the 1950s, it didn't really come into the spotlight until three decades later. Enter two Canadians, Mertice Clark and Bennett Galef, who, in their quest for answers, stumbled upon a peculiar finding in a rodent: the Mongolian gerbil.

Unlike the pear-shaped uterus in humans, a gerbil's uterus has two horns. Picture it like bunk beds in a crowded college dorm – each horn can accommodate two to four foetal gerbils. Now, here's an interesting tidbit about gerbils: Hippocrates, back in the fourth century BCE, asserted that male embryos usually end up on the right, and females on the left. Curious, Clark and Galef put that ancient claim to the test. They checked 1,635 foetuses in 253 female gerbils' uteri through C-section. Surprise, surprise! While Hippocrates wasn't spot-on, he wasn't completely off the mark either. As it turns out, statistically, more males (55 per cent) were found

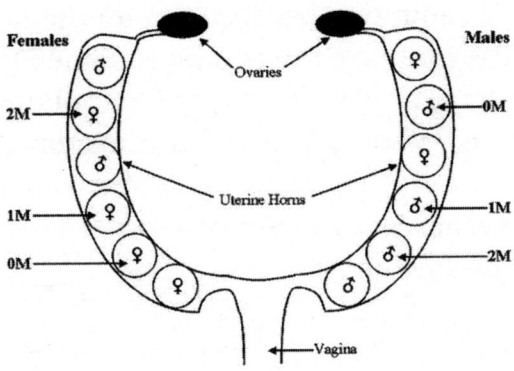

Figure 5.3. A diagram of a two-horned uterus with developing foetuses in a rodent

on the right horn, while more of their female counterparts (58.2 per cent) were found on the left. So there's some truth to this age-old theory.[10]

This gerbil study was just a side project, of course. Clark and Galef's main goal was to investigate what factors influence foetuses to become more masculine or feminine – in this case, they hypothesized that it was the hormones of their in-utero brothers and sisters. The presence of siblings, especially in the immediate neighbourhood, may affect their development. For instance, if a foetus is sandwiched between two brothers (called 2M foetus, where M stands for 'male neighbour'), its growth might differ from a foetus sandwiched between two sisters (known as 2F foetus, where F stands for 'female neighbour').

Indeed, the results were telling – many gender features changed, with 2M males showing a higher level of testosterone compared to 2F males. Now, here's the kicker: just as the Guillette group discovered in alligators with oestrogens, the impact of prenatal exposure to testosterone persisted even after birth in gerbils.[11] Those 2M males, influenced by higher prenatal testosterone levels, turned out to be more aggressive, quicker in mounting females, and less interested in looking after their own pups, in contrast to the 2F males.[12]

Female foetuses weren't exempt from this hormonal influence either. When compared to their counterparts, 2M females (those sitting between two brothers) became masculinized with higher levels of testosterone and lower levels of estradiol. Consequently they were more aggressive and less attractive to males in behavioural tests.[13]

Can something like this happen to humans? Biologist Virpi Lummaa and her collaborators found a smart way to answer the question indirectly. They dug into a trove of family history data in pre-industrial Finland, meticulously kept by church clerks from 1734 to 1888. They then compared the marriage and reproductive histories of both identical and fraternal twins. The findings were striking: women born of opposite-sex twins were more likely to remain single and, if married, had fewer children. This drop in fertility was not observed in any other type of twin.[14] These results were confirmed by another, even larger study in Norway.[15] Now, the question: how could this happen?

Remember how a human embryo is naturally set to develop as female, thanks to the activation of genes like *WNT4* and *RSPO1* – that is, unless the *SRY* gene steps in to flip the script.* Once *SRY* takes charge, it triggers a cascade of gene activity, particularly *SOX9*, boosting the production of androgens like testosterone. But these hormones don't just stick to their own lane – they can seep into neighbouring embryos. This provides a possible reason for how a male twin can masculinize his sister in the womb, subtly altering her development. The primary target for this effect is the brain. As the command centre for the body's nervous and hormonal systems, the brain is where these hormonal crossovers can have a profound impact, shaping everything from behaviour to physiology.

In the mammal world, testosterone, or more broadly androgen, is often used to ramp up masculinity in females. This adaptation is often a response in high-stress situations, kicking in to protect the babies or secure essential resources like food. This plays out beautifully in caribou, for example.

---

* The development of female embryos, initiated by *WNT4* and *RSPO1*, may be more strongly affected by the influence of *SRY* than vice versa. This could account for the observation that male embryos are less impacted by the in-utero environment compared to female embryos.

It's the only deer where females grow antlers too, caused by high levels of androgens.[16] Apparently, being competitive and aggressive is essential for eking out a living among these animals in a tough environment.

This isn't limited to heavyweight champs like caribou either. While they strut around showing off their antlers, something far less impressive yet far more intriguing is happening underground. The European mole – which we met earlier – offers a case of females borrowing from the male playbook, developing traits that blur the lines between sexes. The female mole's gonads are a mixture of functional ovarian tissue and abnormal testicular tissue. During the breeding season, the ovarian part takes the lead. After the eggs are produced and fertilized, and the babies are born, the ovarian tissue fades away. The saved resources and energy from ditching this organ instead go into fuelling the growth of the testicular tissue, cranking out a large amount of testosterone that raises the mom's aggression to the levels needed to compete for food and protect her little ones.[17] Despite her thriving testicular tissue, however, the female mole is technically a female as no sperm is produced.* This case also sends us that same clear message as our humble flea-like insect *Neotrogla* does: biological sex is not necessarily determined by internal sex organs.

In humans, the influence of prenatal testosterone is evident throughout our bodies, and, surprisingly, an easy

---

* A study on the Spanish mole, a close relative of the European mole, has identified two key genes, *CYP17A1* and *FGF9*, as being associated with the development of masculinized tissue. (M. Real, F., Haas, S.A., Franchini, P., Xiong, P., Simakov, O., Kuhl, H., Schöpflin, R., Heller et al., 2020. The mole genome reveals regulatory rearrangements associated with adaptive intersexuality. *Science*, 370(6513), pp. 208–14.)

The Gender Bender

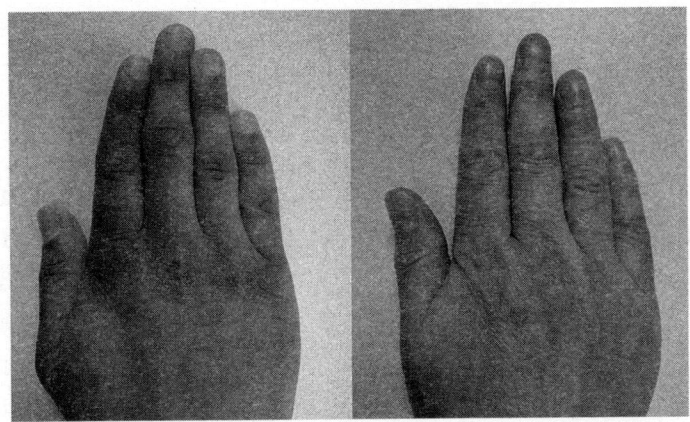

Figure 5.4. Human digit ratio (2D:4D) reflects various levels of prenatal exposure to androgens (A quiz: which hand is more likely from a man and which hand is more likely from a woman?*)

place to observe it is in our fingers. Typically, most men have a shorter index finger (2D, for the second digit) compared to their ring finger (4D, for the fourth digit). Women, on the other hand, tend to have the opposite trend. Studies in mice suggest that a prenatal exposure to high levels of androgens can enhance the growth of the fourth digit.[18] In humans, masculinity or femininity can also be roughly indicated by the 2D:4D ratio, which appears to be linked to a slew of traits – including sexual orientation – related to gender.[19]

• • •

Of course, human gender, especially when it comes to sexual preference, is far more intricate and nuanced than the simplistic masculine/feminine dimension (and their potential hormonal underpinnings) we have talked about so far. It's

---

* Answer: left, man; right, woman.

shaped by the complex interplay of genes, hormones, brain neurons and the environments we live in. (So please don't read too much from your fingers: the data is still far from being conclusive.[20])

First and foremost, genes appear to play a significant role in shaping sexual orientation. According to a comprehensive review by psychologist J. Michael Bailey and colleagues, genetic factors may account for around 32 per cent of the variation in sexual orientation.[21] Gene-mapping studies have flagged several potential hotspots on chromosomes 7, 8, 10, 13, 14 and X, pointing to the trait's complexity – it's likely driven by many genes working together, each exerting a small effect, not just a single 'gay gene'.

One region that keeps popping up in the research is code-named Xq28, located on the long arm of the X chromosome.[22] It's become something of a recurring character in the search for genetic clues. Recently, in one of the largest genome-wide studies to date – spanning nearly half a million participants – scientists identified five genetic loci with promising links to sexual orientation.[23] Despite these leads, the scientific community hasn't yet reached a clear consensus – and for good reason. Sarah Richardson and her research group at Harvard have rightly urged caution, warning against overinterpreting preliminary or inconclusive data.[24]

The same review by Bailey's team also revealed something both surprising and counter-intuitive: non-social environmental factors – disease, nutrition, physical environment – account for about 43 per cent of the variation in sexual orientation, while the social environment – how and where people are raised, and whom they spend time with – accounts for just 25 per cent. These numbers are telling. They challenge the long-held belief that upbringing plays a dominant role. One key 'non-social environmental factor' often goes overlooked: the mother's womb.

Here's one striking example. Scientists have long observed that birth order among biological brothers (but not sisters) affects the likelihood of being gay. Specifically, each additional older brother increases the odds that a younger brother will be gay by about 33–34 per cent.[25] So if the eldest son of a couple has a 2 per cent chance – the lower end of the population average – of being gay, that probability rises to 2.6 per cent, 3.0 per cent, 4.6 per cent and 6.0 per cent for his next four brothers, respectively.[26]

What's especially fascinating is that this effect holds regardless of whether the brothers are raised in the same household. That rules out a purely social explanation. So what's going on?

Research points to the mother's immune system as a key player.[27] When a male foetus develops, it expresses certain Y chromosome-derived proteins – like neuroligins (especially one by the name of NLGN4Y) – that help shape the brain, including regions involved in sexual orientation.[28] Because these proteins come from the father and are foreign to the mother's immune system, her body may respond by producing antibodies. With each successive male pregnancy, her immune system becomes more sensitized – much like how vaccines work – and ramps up its response. These maternal antibodies may subtly alter the development of the next male foetus's brain, nudging it in a direction that increases the likelihood of same-sex attraction later in life.[29] However, exactly how this immune interaction affects the brain is still being worked out.

The birth order effect challenges the common misconception that genetics are destiny. Younger brothers aren't more likely to inherit some 'gay genes' than their older siblings. Instead, it's about how genes interact with their environment – a frontier in biology known as epigenetics. In this case, the mother's body becomes a developmental environment that subtly shapes how certain genes tied to sexual orientation are expressed, particularly in the brain, the hormonal and neural command centre of the body.

This effect also highlights the powerful influence of the prenatal environment in shaping sexual preference – a finding backed by studies in animal models. If sexual orientation is largely set before birth or in early childhood, then the role of socialization – no matter how well intentioned or forceful – is limited. That's why so-called 'conversion therapies' fail so consistently: they're trying to rewrite a script that's already been mostly written. It's also why these practices have been rejected by major professional and scientific bodies, including the American Psychiatric Association and the American Psychological Association.

Why is the prenatal environment so critical to sexual orientation in the first place? To answer this question, let's zero in on the link between hormones and genes in sexual orientation. Although the results in humans are still being debated, studies in mammalian models such as rams show that a foetus exposed to high levels of androgens tends to develop more masculine traits and is usually attracted to females later on. Lower levels of androgens, on the other hand, often result in more feminine traits and attraction to males.[30] The key here, apparently, is how androgens activate specific genes that shape the brain and influence sexual behaviour.[31]

In Mongolian gerbils, exposure to elevated testosterone levels in the womb can rewire the brain's preoptic area, resulting in masculinization of 2M females.[32] Similarly, in sheep, female foetuses exposed to testosterone develop an enlarged preoptic/hypothalamic area, which leads to them taking on more masculine traits.[33] What's especially fascinating is that around 8 per cent of rams exhibit a preference for other rams over ewes. This higher level of same-sex preference is linked to differences in the expression of three specific genes associated with sexual behaviour.[34]

Because many similarities in body systems stem from a shared evolutionary history, studies on lab rodents and farm sheep can provide valuable clues about homosexual behaviour observed across a wide range of mammals – elephants, antelopes, whales, seals, dolphins, macaques, baboons, chimpanzees, bonobos and more. Such behaviour is also seen in over 130 species of birds like penguins, geese and terns, as well as in invertebrates such as insects and octopuses, a total of more than 1,500 species and counting.[35] In fact, the more scientists look for, the more cases they find, indicating just how widespread this is.

Excitement aside, the key point from what we already know is that the biological construct of gender is not just about behaviour, physiology or morphology; it is rooted in genes, hormones and the organization of crucial brain areas.* With social construction added as a new dimension, human gender emerges as a richer and more complex phenomenon. Yet we can remain hopeful that a collaborative effort between biologists and social scientists will lead to a clearer, more objective understanding.

. . .

---

* While 'gender' here refers to a collection of biological features across all body systems and at all levels from the cell to the entire body, 'gender role' refers specifically to behaviours as they are shown.

We've looked at *how* hormones and birth order can influence gender orientation. Now we're diving into an even bigger mystery: *why* human sexuality is so dazzlingly diverse. Consider same-sex preference – it's a trait that consistently shows up across cultures, time periods and social systems. Depending on how it's measured, surveys find it in the range of 2–10 per cent of the population.[36] So why is it so widespread?

As we've seen, same-sex behaviour isn't a uniquely human phenomenon – it's a recurring theme in nature, observed across a vast array of animals, from vertebrates to invertebrates. So the answer, as with so many mysteries of life, likely lies in evolution. These behaviours, far from being anomalies, may provide adaptive benefits that have allowed them to persist for aeons.

Dozens of evolutionary hypotheses have been tossed into the ring, suggesting that same-sex behaviour could help forge alliances, reduce conflict, or boost an individual's overall appeal to potential mates.[37] But the elephant in the room remains: how do individuals with little or no fertility manage to keep their genetic legacy alive?

One sweeping answer, as we explored in the previous chapter, lies in kin selection. You don't have to reproduce yourself to ensure your genes' survival; helping your relatives thrive may achieve the same result. This can make the intersex spectrum an adaptive strategy, one that endures across a wide range of animal species from social insects to birds and mammals. It's only logical that kin selection may also have played a major role in the evolution of same-sex preference.

Humans have long relied on kin selection to pass along shared DNA. That's why we instinctively invest time and resources in our relatives – our children, grandchildren, nieces, nephews and cousins. Traditional social structures, like large families and close-knit clans, have served as fertile ground for kin selection, fostering cooperation and mutual support for generations.

Kin selection can manifest in puzzling cultural practices, even ones that seem counter-intuitive at first glance. Take, for example, the well-documented preference for sons in many Asian cultures. In some traditional farming societies of Southeast Asia, however, the opposite occurred: a young mother might deliberately kill her firstborn son, believing a daughter would be more valuable. Why? Because daughters, in those communities, played a crucial role in helping raise the mother's future children, increasing her overall reproductive success.[38] This stark and unsettling practice illustrates how kin selection could make the mother's infanticide adaptive: she traded one part of her fitness (her newborn son) for another (her future children).

Kin selection is a compelling explanation for the evolution of same-sex preference, but it doesn't exclude other possibilities – as you've probably realized by now, biology is rarely that straightforward. Indeed, in 2024 a bold alternative to kin selection emerged.[39]

Analyzing UK Biobank data,* University of Michigan biologists Siliang Song and Jianzhi Zhang found that bisexual and gay men tend to produce fewer children than heterosexual men, with bisexual men faring slightly better. No surprises there. But they also discovered that bisexual men, carrying the genetic markers of both heterosexual and homosexual genes,†

---

* UK Biobank is a long-term research project with anonymous genetic, lifestyle and health data from half a million people in the UK.
† Given the absence of any definitively identified genes linked to same-sex preference, the researchers employed genetic markers (known DNA fragments) associated with sexual preferences as proxies for these genes.

are more prone to risk-taking – a trait with huge pay-offs in pre-modern societies: bold men had more chances to 'sow their oats', spreading their genes widely, especially in polygynous cultures.[40]

The genes linked to male bisexuality, then, may not be just about sexual preference but also about reproductive success in environments that favoured risk-takers such as violent adventurers. (Think of figures like Moulay Ismael the Bloodthirsty, who reportedly fathered 888 children, or Genghis Khan's Mongols and the Spanish conquistadors, whose genetic legacies are legendary.) This might be one reason why same-sex preferences exist in human populations,[41] as well as other mammalian species. Could some of history's boldest conquerors have carried these genes, fuelling their audacious exploits and prolific offspring? It's a hypothesis worth investigating.

Song and Zhang's findings are still preliminary, underscoring how much remains to be uncovered about the evolution of same-sex behaviour, even though it is widespread across a vast array of animal species. It's likely that additional, currently unknown mechanisms also contribute to the remarkable diversity of sexual behaviours observed in nature.

• • •

Now that we've explored the biological foundations of gender and its adaptive role in the rich diversity of sexual preference, it's time to tackle the chapter's next big question: why do some animals switch genders? What is the evolutionary logic behind these transformations?

Let's set the stage with a thought experiment: imagine you have complete freedom to manipulate your sex or gender identity. How would you maximize your Darwinian

The Gender Bender

fitness? If you're taking cues from clownfish or swamp eels, you might decide to switch sexes when the situation calls for it. But this isn't some simple swap – it's a full-scale internal reconstruction. You'd need to rewire your brain, flood your body with new hormones, dismantle outdated body parts, grow entirely new tissues and organs, and overhaul your appearance and behaviour. It's not just a transformation – it's a total renovation, like hiring an army of contractors to gut your house and rebuild it from scratch. It's expensive, time-consuming, and once you're in there's no easy way out – backing out is either impossible or will cost you a fortune.

Indeed, sex change is no casual affair in nature – it's usually slow, permanent and inflexible. The hefty cost makes it worthwhile only under a very specific and predictable set of conditions where switching sexes gives a clear Darwinian advantage. Imagine you're a clownfish. You spend years bulking up and climbing the social ladder as a lowly and humble male, just waiting for your moment to shine as the top female. But right after you've made the big switch, the game changes and, suddenly, being male is the better bet.* All that time, energy and effort? Down the drain. What a bummer!

These drawbacks of sex change are a nightmare for animals that need to quickly switch gears in an unpredictable world. It's no wonder most species steer clear of this evolutionary detour. So what's the secret to surviving and thriving in a fast-changing environment? The answer lies in one word: gender.

---

* The reduced fitness observed in male clownfish appears to be attributable to their high mortality risk from predation when venturing beyond their home reefs. Consequently, remaining within the natal reef and awaiting the opportunity to undergo sex change represents a better strategy compared to dispersing in search of external mating opportunities.

Instead of going through the complex, high-stakes process of changing sex, bending gender offers a simpler, faster and largely cheaper option. It's like upgrading your home. You can start small, like rearranging the furniture at no real cost. Or, if that's not cutting it, you can go big, all the way to tearing down walls and redoing the decor. Bending gender, likewise, gives animals a range of options to boost their fitness, but it's way more rapid, efficient, flexible and usually far less costly than a full-on sex change. That may explain why it's so common in animals.

As you can see, if sex divides animals into producers of large versus small gametes, gender does the opposite – it brings them back together. The whole point of gender development is to give animals the flexibility to swap, share or mix male and female roles depending on what the situation calls for. In that sense, gender isn't just a label – it's a versatile survival tool, tailored to help each individual navigate the social and ecological maze it's born into. Thus, for most animals, there's no need to envy the flashier trick of changing sex. Bending gender gets the job done just fine – and usually it's faster, cheaper and way less dramatic.

So what's nature's ultimate lesson about gender? Flexibility.* The ability to custom-build bodies and behaviours to fit whatever survival and reproductive challenges the environment throws your way. Consider our old friend, the jacana. With chicks constantly at risk from predators above and below, it makes more sense for jacana dads to double down on parenting than to roll the dice chasing new mates.

---

* In this context, gender is conceptualized as a continuous trait, rather than a discrete reproductive polymorphism, as proposed by some biologists – most notably Joan Roughgarden in her influential 2009 work *Evolution's Rainbow* (Berkeley, CA: University of California Press).

The Gender Bender

Think of it like finishing one solid project instead of juggling a bunch of half-baked ones that never get off the ground. As the dads settle into their full-time caregiver roles, the moms take centre stage – strutting, squabbling and racking up feathery flings in ways usually reserved for males in most bird species. The result? A total flip of the usual gender script, courtesy of evolution's favourite tool: adaptability.

Something similar might be nudging seahorses – and other gender-benders – down the same path. Seahorses live a laid-back, low-key life, clinging to sea grass or coral and barely moving around. They're not exactly social butterflies either, often scattered in small numbers and so well camouflaged that they can fool even the sharpest-eyed scientists while hiding in plain sight.[42] For male seahorses, chasing after multiple mates just isn't worth the hassle. Their lifestyle doesn't lend them to racking up reproductive points that way. Instead, their best shot at passing on their genes is to commit fully to what they can control: protecting and nurturing their young.

Figure 5.5. A female seahorse (right) delivering her eggs to the brood pouch of a male (left)

These examples reassure us of what we already know about complex animals. As sperm and eggs are tiny and helpless in life's two forms of existence, the body – or, as we brought up in Chapter 3, Weismann's soma or Dawkins's vehicle – has to assume all survival functions leading to the successful delivery of the genetic package to the next generation. That means every part of the male or female body – whether it's about shape, structure, physiology, hormones or brain wiring – is designed and can be tweaked ultimately to serve this evolutionary plan. So, *gender in biology should be seen as a broad collection of traits that lead to the behaviours – gender roles – the body assumes*. Gender reversal shows just how flexible gender can be bent to meet socioecological demands.

With that said, full-on gender reversals are rare in the animal kingdom – think of them as nature's version of going all-in at the poker table, a risky move only pulled when the odds are truly exceptional. But partial reversals? Those are everywhere. Animals routinely fine-tune a whole range of gender-related traits – size, metabolism, hormones, brain wiring, behaviour – you name it. They adjust on the fly, responding to whatever curveballs their environment throws at them. This kind of flexibility is a serious evolutionary superpower. It boosts their Darwinian fitness and helps them roll with the punches in a world that's always changing. So gender isn't just a spectrum – it's a shape-shifting, ever-evolving state, constantly in motion.

• • •

Human gender roles are uniquely characterized by their remarkable flexibility, a feature unparalleled among animals and rooted in humanity's dynamic and rapidly evolving environment – especially sociocultural environment. This biological and cultural adaptability explains why gender varies widely across societies, even as the foundational categories of 'men' and 'women' remain universal. For example, as widely documented in both scholarly and popular works, gender roles in Greco-Roman societies looked quite different from those in other ancient civilizations – like China, India or the Inca. These contrasts reflect just how deeply and profoundly culture shapes gender.

Instances of partial gender role reversal are abundant throughout human history. The role of the warrior, often seen as a domain for men, has long included formidable women. Legendary Amazon princesses have real-life counterparts in historical figures such as Boudica, Joan of Arc, Sikh warrior Mai Bhago and the fighters of the Dahomey kingdom (modern-day Benin). Today, in industrialized nations, the sight of women in military uniforms hardly raises an eyebrow.* Women warriors are a striking example of how human gender roles shift, especially in response to socioeconomic conditions.[43]

An equally fascinating example comes from dynastic China: the practice of foot-binding. From the late thirteenth to the early twentieth century, bound feet were seen as the height of femininity among the upper class – a mark of beauty, delicacy and privilege. But for poorer families, survival trumped fashion. Women were essential to farm work, and feet too fragile for the fields were a luxury they simply couldn't afford. In homes where every pair of hands – and feet – counted, having a wife who could work mattered far more than keeping up with elite ideals. This sharp divide

---

* During the founding of Islam, several of Prophet Muhammad's wives provided advice to and fought bravely with him.

between the rich and the poor illustrates just how much gender can be shaped by economic necessity as much as by cultural ideals – and therefore speaks to gender's inherent flexibility where the human species is concerned.[44]

More familiar to Westerners, the medieval Western perspective on gender was likewise forged by cultural factors such as socioeconomic conditions and Christian beliefs. The division of labour along gender lines was designed, at least in part, to efficiently manage economic activities and domestic responsibilities within the nuclear family. Men were often heralded as 'breadwinners' because they usually earned the income that supported the family. Meanwhile, the vital role women played in maintaining the household – especially in raising children – was often downplayed, as their work lacked monetized value, essential for status and recognition in these societies.

The advent of the machine age reduced the significance of physical strength, a trait historically linked to men, and profoundly disrupted traditional gender roles in medieval Western societies. It is no coincidence that the suffragette movement gained momentum in the latter half of the nineteenth century – as if Western women had suddenly become aware of their rights after millennia of marginalization. The Second World War helped to catalyze this shift, as women, particularly in the United States, demonstrated their competence in operating machinery in factories. Much as the Agricultural Revolution brought an end to the era of man the hunter, the Industrial Revolution heralded the decline of man the breadwinner.*

---

* That being said, gender discrimination has not completely disappeared, even in democratic nations.

The Gender Bender

In today's Western economies, brute strength has taken a back seat to brainpower. We've entered the age of ideas, innovation – and now, AI. Creativity and connection are the new currency, and women have stepped confidently into the spotlight. No longer boxed into traditional roles, they've become indispensable across every corner of society. From factory floors to executive suites, farmland to Wall Street, classrooms to operating rooms, women are everywhere. They're building roads, leading companies, and yes, even launching into space. In the US alone, 15 million women are running single-parent households, skilfully juggling the roles of both mom and dad. This massive shift in economic and social dynamics has made the gender roles of the boomer era look about as modern as treating a fever with medieval bloodletting.

Despite rapid and sweeping socioeconomic changes and the incredible diversity across cultures, however, men and women remain the two primary genders today. This persistence highlights the deep biological roots of gender, particularly its role in direct reproduction. It also sheds light on behaviours often observed in children, such as boys favouring trucks and girls gravitating towards dolls, as well as cognitive differences like women outperforming men in verbal fluency and vocabulary size.[45] Such gender differences, as anthropologist Agustín Fuentes explores in his 2025 book *Sex Is a Spectrum*, extend beyond behaviour to encompass morphology, physiology, endocrinology, neurobiology, psychology, genetics and more. They reflect the adaptive value of gender in our evolutionary history, where biology and culture intertwined to shape survival strategies. Today, this interplay continues to influence how we navigate the fast-changing world.

Even so, our understanding of gender variations is still rudimentary, with much awaiting to be discovered, especially in terms of evolutionary significance. This demands both open-mindedness and restraint – avoiding hasty conclusions and radical claims based on what we know, or worse, what we *think* we know.

• • •

While there's still much to learn, one thing is already clear: sex and gender diversity are part of our biological reality. The real challenge lies in how society talks about them – especially when those conversations are dominated by the majority. This is particularly true in democratic societies, where majority opinion often sets the tone. So how can a minority – especially those who dare to love differently – be truly understood, accepted and respected by a heterosexual majority? That question has fuelled social struggles across many nations.

In the United States, open conversations about homosexuality didn't really gain traction until popular culture stepped in. Films like *Philadelphia* (1993) and TV shows like *Will & Grace* (1998) brought the challenges of the gay community into mainstream living rooms. Around the same time, scientists like Dean Hamer at the National Cancer Institute helped shape public understanding with academic and popular works exploring the genetic roots of sexual orientation. Together, these cultural and scientific shifts began to chip away at long-standing stigma – and opened the door for deeper dialogue. Before long, Congress enacted the controversial 'Don't ask, don't tell' policy, allowing gay, lesbian and bisexual individuals to serve in the military – so long as they kept their identities hidden. It took over two decades after the Supreme Court's landmark ruling in *Obergefell v. Hodges* for same-sex couples to secure full legal recognition for marriage and adoption.

You might wonder why progress in recognizing such a basic human right has been so painfully slow. Alexis de

Tocqueville's idea of 'the tyranny of the majority' offers a clue. In a democracy, the will of the majority can quietly – yet powerfully – suppress the needs of minorities, especially when the majority is unaware of their struggles. Since most people identify as heterosexual – considered the default or 'norm' – there's often little urgency to question the status quo. That quiet indifference has been one of the biggest roadblocks to meaningful change.

The numbers tell the story: since the 1970s (up to the time of writing), anti-same-sex marriage or anti-civil-union propositions have passed more than fifty-nine times. Meanwhile, efforts to legalize these unions have been voted down fifty-nine times since the 1990s. This stark imbalance underscores the persistent challenge of securing minority rights in the face of deeply ingrained majority views.[46]

If you, like me, believe that most voters are fundamentally kind-hearted, then the second reason – lack of scientific understanding – becomes key. For a long time, sexual preference was wrongly viewed as a lifestyle choice, something personal and changeable. It's only in recent decades that science has illuminated the deep biological roots of sexual orientation. In this way, science acts as a liberator, dismantling false beliefs and freeing us from the weight of unfounded prejudices. Objective knowledge, as always, is the first step towards understanding, compassion, justice and, ultimately, reconciliation of truth-loving people with different ideologies.

Indeed, as we've seen about the broad spectrum of sex and gender, there is no single best way to build Dawkins's vehicle for our body as male or female. This will become even clearer in the next section, where we'll explore how the two forms of existence can be tweaked in a variety of ways to pass on genes across generations.

Part III
THE STRUGGLE

6

RED IN TOOTH AND CLAW
Mate Competition

On 28 April 1789, chaos erupted aboard the HMS *Bounty* – a scrappy, 91-foot, three-masted British cutter, the runt of the Royal Navy's fleet. Drifting near Tofua (now part of Tonga), the little ship was carrying out a scheme cooked up by Joseph Banks, president of the Royal Society.* Banks's big idea was simple enough: sail to Tahiti, scoop up breadfruit plants and ferry them off to the West Indies as cheap food for enslaved plantation workers. Now, jam-packed with 1,015 fragile seedlings in pots covering the decks, the *Bounty* must have looked less like a warship and more like a floating botanical nursery.[1]

But beneath decks, trouble was simmering. The sailors were fed up with Captain William Bligh's tyrannical command – his explosive temper, brutal punishments and relentless discipline. Finally, driven by anger (and fuelled by more than a little rum), Master's Mate Fletcher Christian and his band of mutineers snapped. They stormed the deck, seized the *Bounty* and wrested control from Bligh. The humiliated captain and eighteen loyal crewmen were forced on to a pathetic little lifeboat, barely 23 feet long and 6.6 feet

---

* It was (and still is) Britain's prominent natural science academy for 'improving Natural Knowledge'.

wide, stocked with just five days' worth of supplies. As the rope was sliced, the abandoned men watched helplessly, drifting away into the endless, unforgiving sea.*

The mutineers knew exactly what they were risking – this kind of treason came with one guarantee: a noose. Hunted by the ever-watchful British Navy,† they were desperate to vanish off the map. After a string of failed hideouts and close calls, they turned their sights to Pitcairn Island – a lonely speck in the vast Pacific, mischarted and nearly impossible to find. It was the perfect place to disappear. Onboard for this last-ditch escape were nine mutineers rolling the dice on a life beyond the Crown's reach: Fletcher Christian, Edward Young, William Brown, John Mills, William McCoy, Isaac Martin, Matthew Quintal, John Williams and John Adams. Alongside them were eighteen Tahitians – six men and twelve women, mostly abducted – dragged into a high-stakes exile they never chose.‡

When they finally stepped on to Pitcairn's rugged shores in early 1790, relief washed over them. The island – just 2 miles long and 1 mile wide – felt like paradise. The climate was mild, the place teemed with food – breadfruit, coconuts, fish and wild birds – and, best of all, the jagged coastline made it nearly impossible for ships to land. It was the perfect hiding spot. Determined to erase any trace of their mutiny,

---

* Bligh was given basic tools for navigation including a compass, a quadrant and an unreliable sextant. Amazingly, the small boat drifted for 3,618 nautical miles in forty-eight days and made it to Timor, then part of the Dutch East Indies. Eventually, twelve of the nineteen men returned to England on 13 March 1790.
† The Royal Navy did send a 24-gun warship, HMS *Pandora*, to capture the mutineers, but it wrecked on the Great Barrier Reef before finding those hiding on Pitcairn.
‡ Only some of the names of the Tahitians are known including Fletcher Christian's wife Maimiti (also known as Isabel).

the men gutted, dismantled and finally set the *Bounty* ablaze, watching their old life go up in flames as they set about building a new one.

But this lush island paradise didn't come with a guarantee of happily ever after. Far from it. As it turned out, finding the perfect hideout was the easy part – learning to live with each other was the real battle.

The trouble centred around the Tahitian women, who became flashpoints for major conflicts. The Englishmen treated these women like property, trading them from one 'husband' to the next. This twisted practice of 'sharing' stirred jealousy, rivalry and outright hostility among the settlers. For three tense years the conflict simmered until, in September 1793, it boiled over into bloodshed. In the chaos, Christian, Williams, Brown, Mills and Martin were killed. But the violence didn't stop there. Within a year, all six Tahitian men were dead too. By 1799, McCoy had taken his own life, and Quintal had been murdered. Even then, peace remained out of reach. It wasn't until a year later, after Young's death, that things finally settled, leaving the stocky, five-foot-five Adams as the last man standing among nine Tahitian women and the nineteen children born on the island.

With no male rivals and a de facto monopoly on all the women, Adams assumed the role of preacher, wielding the lone Bible salvaged from the *Bounty* like a sacred sceptre. On this remote enclave, the man ruled the fiefdom by his word, and his teachings became their gospel.

The Pitcairn colony remained hidden from the outside world for eighteen years until 1808, when the American sealer *Topaz* stumbled upon it. Captain Mayhew Folger found thirty-five souls eking out a life on the remote island. At the time, Britain was embroiled in war with Napoleon's France, and the pursuit of the mutineers had lost its urgency. Adams, once a hunted man, was eventually granted amnesty. Official records list six of his children – Thursday October Christian, Sarah Christian, Elizabeth Christian, Catherine Christian, Hannah Adams and Edward Young – but it's likely he fathered several more with the Tahiti women before he passed away at the age of sixty-two in 1829.*

This dramatic and violent story about the mutineers of Pitcairn Island reinforces Aristotle's assertion that, in lawless societies, men are 'the most unholy and the most savage of animals, and the most full of lust and gluttony'.[2] And for Thomas Hobbes, a society becomes a battleground for 'war of every man against every man'.

Aristotle and Hobbes, perhaps without meaning to, were on to something when they spoke of 'man' rather than 'woman'. But why are men, specifically, prone to commit such violence in the absence of governance? The answer lies in biology: the male body – Dawkins's gene-shuffling vehicle – is fine-tuned for one thing above all else: mate competition. This relentless struggle for reproductive success is at the core of sexual selection, one of evolution's driving forces. Even under extreme conditions, the mutineers' story becomes a kind of pressure cooker – a place where 'maleness' boils to the surface in one particular species in the animal kingdom. But why are males so … *male*? To find out, let's turn to nature – it's got plenty to say on the matter.

. . .

---

* At the time when Mayhew landed on Pitcairn, thirty-five people were living on the island. Assuming all the nine Tahiti women survived, Adams might have already sired six more children by then.

Red in Tooth and Claw

As we saw in Chapter 3, the evolution of sex began with two epic wars. First came the battle between gametes – a microscopic showdown that ended in a truce: small gametes became sperm, large ones became eggs, and their producers became males and females. But that deal sparked the next big conflict – this time within each sex. Males began locking horns with other males, and females squared off with other females. Two ruthless civil wars, all in the name of reproduction. At the heart of this conflict lies the real question: why do members of the *same* sex turn on each other?

For males, the numbers tell a frustrating story. Sure, sperm are abundant, eggs are rare – but it only takes one of each to spark a new life. And in most cases, the ratio of males to females stays balanced (for reasons we'll dive into later). As long as food and resources aren't stretched too thin, a male's biggest obstacle in his path to reproductive success often isn't the environment, bad luck or even female choice – it's other males, all equally desperate for a shot at passing on their genes.

Now, imagine we were dropped into the mutineer colony on Pitcairn Island. We may not be much different from a bunch of lawless primates. Take Adams, the island's alpha male. He might flex his dominance by snatching a banana from a female now and then, but he wouldn't be worried about them knowing his status. His real enemies, as far as he was concerned, were the other men, not the women. His reproductive success wasn't measured against the females – it was measured against his male rivals. And in a brutal twist, Adams's winning strategy was wiping out all the competition – which is exactly what happened on Pitcairn.

But what worked for Adams wouldn't quite apply if he were instead a she – an imaginary Eve. In many animals, including us, females don't go to such extremes to eliminate their same-sex rivals, and we'll get into why soon. It's a bit like most competitive sports – men and women both compete, but they chase different trophies, often by playing with different rules and strategies.

Imagine the wild world of mate competition, and you'd probably conjure up scenes like these: bull moose lock their massive antlers and go head to head in a battle royal, all for the chance to woo a harem of females. Roosters aren't any gentler – they peck and stab each other with sharp beaks and spurs, fighting for top billing in the pecking order. Male lizards are all about turf wars, chasing off any intruder with the intensity of a bouncer at an exclusive club. Meanwhile, male frogs cling to the backs of females like their lives depend on it, kicking away rival suitors with all the tenacity of a determined hitchhiker. Male fish dart at intruders, asserting their dominance like tiny, scaly warriors.

While there are exceptions – like hyenas, jacanas and seahorses, where the females are the ones throwing punches – these cases are truly the oddballs. So why do males tend to be the more violent ones? The answer lies in the first 'battle of the sexes', the evolutionary showdown that split sperm from egg. That ancient clash set the stage for all the hostility that followed.

When resources are plentiful, eggs can thrive on their own in theory. Just look at the rare but real cases of virgin births in vertebrates – California condors, Komodo dragons and a few other evolutionary mavericks have pulled it off. Sperm, however, haven't been so lucky. By opting for the 'small and plentiful' strategy, they gave up any chance of independent survival to specialize in a parasitic lifestyle. That means sperm have exactly one job: find an egg and fuse with it.

But here's the catch – sperm vastly outnumber eggs. That simple imbalance gives males far more reproductive potential than females. Imagine males as delivery trucks overflowing with packages (sperm) and females as a limited number of households in a neighbourhood, each receiving just one on the doorstep. With way more packages than available homes, male success depends on speed, persistence and sheer numbers. So, all else being equal, it pays for males to be more aggressive in competing for access to females, because their reproductive success depends on it.

Figure 6.1. Evolved weapons in fish, amphibians and reptiles. Some of these specialized traits pull double duty, helping individuals compete for mates and charm them too

Caught in a high-stakes game where the odds are stacked against them, the haploid sperm, diminished and apparently helpless, react by outsourcing most functionality to their other form of existence – the diploid male body – to survive and fight battles against rivals. That's why male animals often face a tougher physical test compared to females. Under this intense pressure, males have evolved a slew of traits to excel in mate competition, including everything from elevated levels of aggression to menacing weapons and bulky body armour.

As early as the nineteenth century, Darwin speculated that certain male traits evolved through mate competition. In *The Descent of Man* he wrote about 'the greater size, strength, and pugnacity of the male, his weapons of offence or means of defence against rivals, his gaudy colouring and various ornaments, his power of song, and other such characters'. Recent studies have confirmed Darwin's ideas. For example, canines in primates and horns or antlers in ungulates (hoofed mammals) have evolved more for fighting male rivals than for defending against predators.[3]

Let's bring up a biological principle called Bateman's rule, which suggests that males can maximize their fitness by pursuing females, while females do so by seeking resources. This rule doesn't consider social and ecological factors, and so it has many exceptions and isn't universally applicable,* but it remains relevant in many animal species. For males,

---

* A recent study showed that there are several critical errors in Bateman's 1948 article, especially about female reproductive strategies in *Drosophila melanogaster*', (Gowaty, P.A., Kim, Y.K. and Anderson, W.W., 2012. No evidence of sexual selection in a repetition of Bateman's classic study of *Drosophila melanogaster*. *Proceedings of the National Academy of Sciences*, 109(29), pp.11740–45.)

physical strength often provides a competitive edge in securing access to females for mating opportunities.

Male weaponry – horns, antlers, tusks – might steal the spotlight in some species, but nature has plenty of other tricks up its sleeve when it comes to mating battles. More often than not, males go for brute size rather than built-in armoury, simply growing bigger than females. And in all known cases in animals, they pull off this growth spurt by tweaking a gene called *dmrt* – nature's little genetic switch for size advantage.[4] The result? A glaring size gap between the sexes, a classic hallmark of sexual selection.[5] Just look at mammals where mate competition has pushed males to the extreme in many species: male lions can double the size of females, male gorillas can triple it, and male elephant seals can outweigh their female counterparts by a factor of ten, turning their bodies into sheer battering rams for reproductive success.*

Big bodies and killer weapons don't just pop up out of nowhere. It's not as simple as slapping on some horns or fangs and calling it a day; the whole male body has to step up its game to become a top-notch Dawkins's male vehicle. That means revving up the metabolism, flooding the system with hormones and readapting the body for combat missions. But all these perks come at a price. Burning through tons of energy and resources is just the start – throwing down in physical fights is a great way to rack up some severe injuries

---

* A survey at the order level shows that instances of male-biased, female-biased and sexually monomorphic body size account for 45 per cent, 39 per cent, and 16 per cent of cases, respectively. (Tombak, K.J., Hex, S.B. and Rubenstein, D.I., 2024. New estimates indicate that males are not larger than females in most mammal species. *Nature Communications*, 15(1), p. 1872.) However, in no mammalian species do females exceed males in body size by multiple-fold, a pattern commonly observed in certain invertebrates and fish.

On the Origin of Sex

Figure 6.2. Two rattlesnakes trying to subdue each other with physical strength

or even punch your own ticket. For creatures packing serious heat like horns, antlers, tusks or venom, the real trick is figuring out how to win a battle without ending up in worse shape than their opponent.

The solution to this kind of high-stakes conflict is to evolve a set of ground rules. Take venomous snakes like vipers and rattlesnakes, for example. If they started biting each other left and right, it'd be game over for both the combatants.* So they've evolved a species-wide gentleman's rule: instead of going for the lethal bite, they engage in a lengthy, exhausting tug-of-war, where one tries to pin down the other's upper body like a couple of overgrown noodle wrestlers – until one of them gives up.[6]

---

* They can digest and detoxify the venom, but a direct injection like biting would still harm them.

In the same vein, male giraffes have a unique approach to conflict resolution – they duke it out by swinging their necks like wrecking balls, rather than risking lethal kicks that could take down a lion. You'd think that massive elephant bulls would be using their tusks for all-out war, but instead they lock tusks and push, more like sumo wrestlers than swordfighters, keeping the bloodshed to a minimum. Even male narwhals, armed with their impressive 10-foot-long javelin tusks, prefer to keep things civil. Instead of going for the kill, they engage in what can only be described as chivalrous fencing matches when competing for mates. They may be sparring for love, but they're not about to turn it into a deadly duel. *Touché!*

If you're looking closer to home for this phenomenon, you can consider the red deer. A detailed breakdown from a multi-year study by biologist Tim Clutton-Brock and his colleagues showed that bulls often resolve conflicts through roaring. If that doesn't suffice, they then engage in a ritual called parallel walking, striding side by side so each can size up the other – literally. It's a tense showdown where body size does the talking. Remarkably, 84 per cent of clashes are settled without physical contact at this stage. Only 16 per cent of conflicts escalate to physical sparring with antlers, and even then the fights are relatively restrained, with just 6 per cent of males sustaining permanent injuries.[7]

If you're like me and tend to see wild animals as fuzzy little warriors of lovey-dovey cuteness, you might think they're playing fair out of the goodness of their hearts. But the truth is far less romantic: it's all about practicality. Animals follow these unwritten rules because it's the smartest way to settle scores without breaking the bank. Even the loser gets something out of it – he (and his genes) lives to fight another day, maybe against a weaker rival or when the stars align just right. Mutual destruction is a lose-lose situation, disfavoured by evolution. That's why you'll see more low-stakes face-offs – like singing, roaring and hissing – than full-blown brawls. It's like a game of poker where nobody really wants to go all in.

Similarly, most human tribal communities manage to get along just fine most of the time, which led Jean-Jacques Rousseau to idealize the concept of 'noble savages'. He was convinced that culture was the bad apple that spoiled our good nature and didn't believe that civilized folks could keep their act together without a little rulebook. In his 1762 work *The Social Contract*, Rousseau, taking a cue from the ever-pessimistic Thomas Hobbes, argued that without some top-down control, we'd all be at each other's throats faster than you can say 'chaos'.

What made Rousseau's 'noble savages' live together peacefully? This question nudges us towards the idea that a social contract – or at least a rough draft of it from evolution – might have been around long before philosophers started putting pen to paper.* It also makes us rethink the notion that animal mate competition is just 'red in tooth and claw' at its most extreme. If male rattlesnakes, narwhals and red deer can make it through the mating game without tearing each other apart, maybe all those great scientists since Darwin have been a little too laser-focused on competition. Could the 'live and let live' scenes we often witness in nature hint at something more than just cut-throat competition? The best way to find out is simple: experiment.

---

* The social contract refers to an agreement within a society to cooperate for collective benefits, often at the cost of certain individual freedoms in exchange for state protection. Initially a philosophical concept, its parallels in social animals have been increasingly recognized in recent decades.

My research team and I decided to put this idea to the test with guppies in the lab. We placed two males in a tank with two females, each stationed at opposite ends. You might expect that the two males would battle non-stop until one conceded defeat, but the reality was more complicated. Yes, they fought a lot, but they also knew when to call a truce to flirt with the females – because let's face it, you can't woo a lady while you're in the middle of a brawl.

According to Bateman's rule, males are expected to go relentlessly for getting egg-producing females as their ultimate prize. But in our study, when male guppies weren't busy squabbling over mates, they actually took turns courting the females. So, beyond the rivalry, there was a surprising level of cooperation – though it might have been born out of necessity rather than chivalry. This kind of begrudging teamwork is what economists call cooperative bargaining, and it's hard not to see it as an early blueprint for a social contract.

The evolved rules of engagement come with a hitch: they can't always stop rule-breakers from taking advantage. In the animal world there's no one to play referee, so casualties do happen once in a while, and they often go unpunished. That's why you occasionally see animals killing or even cannibalizing each other. Infanticide of a rival's baby, for example, is surprisingly common, especially among carnivores and primates, often driven by the brutal logic of mate competition. It's nature's dark side, where the 'live and let live' deal sometimes gives way to a more ruthless survival strategy.

In Indigenous societies, violence within and between tribes is generally more common than in so-called civilized societies with written laws (we'll return to this later). Rousseau – and other thinkers like Hobbes and Kant – were on to something when they advocated for governments to enforce social contracts, reducing conflict and promoting peaceful living. The rule of law and order in societies didn't just lower the frequency of violence; it paved the way for human cooperation on a massive scale, which enabled megaprojects like the pyramids, the Great Wall and St Peter's Basilica – feats no tribal society could dream of pulling off – as well as today's multinational treaties and the United Nations. Apparently, Rousseau's cynicism about human culture missed the point by a mile. He didn't see that civilization has actually made society more peaceful and cooperation more robust, allowing us to work together on an ever-larger scale.

...

Now let's refocus on mate competition. It doesn't end when sperm is released. In fact, that's just the beginning. Between the two forms of existence, males – the sperm-producers, carriers and deliverers – serve only as middlemen in the grand scheme of evolution. The real players are the sperm. They are the principal agents in the ultimate business of passing down genes, while the males are just the couriers trying to make sure the package gets to its destination.

In most animal species with internal fertilization – think insects, sharks, reptiles, birds and mammals – females don't exactly play the loyalty card. Whether openly or on the sly, they often mate with multiple males, as a means of mate choice or diversifying the genetic make-up of their own offspring. In a way, this makes delivering sperm a bit like counting your chickens before they hatch. The battle continues even after the sperm is inside the female's body, where it's a race against both time and other males' sperm to reach the eggs.

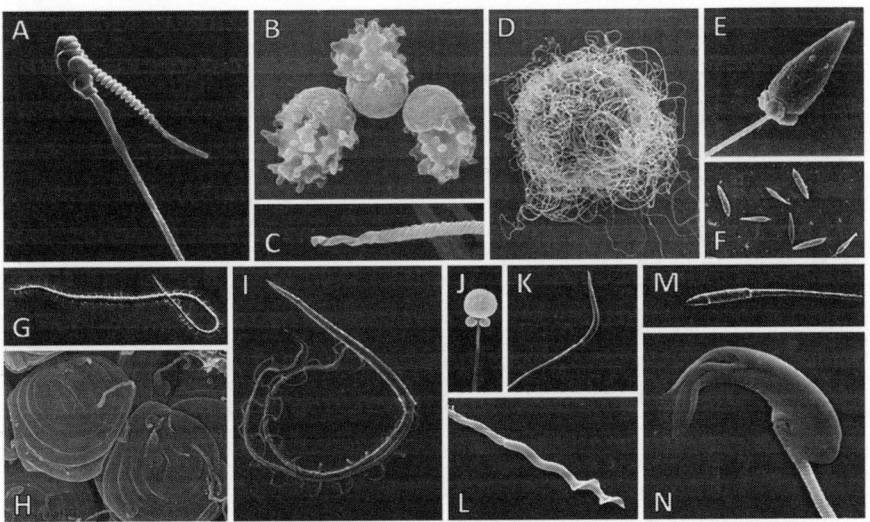

Figure 6.3. Adaptive diversity in sperm morphology in various animals. Scanning microscopic image of A. *Macrobiotus* (L. Rebecchi), B. *Caenorhabditis elegans* (T. Roberts), C. *Mytilocypris mytiloides* (R. Matzke-Karasz), D. *Drosophila bifurca* (R. Dallai), E. *Patinopecten yessoensis* (Q. Li), F. *Iporangaia pustulosa* (J. Moya), G. *Trialeurodes vaporariorum* (R. Dallai), H. *Allacma fusca* (R. Dallai), I. *Colostethus marchesianus* (A.C. Veiga-Menoncello), J. *Paralichthys olivaceus* (Y.Z. Zhang), K. *Gopherus agassizii* (L. Liaw), L. *Passer domesticus* (R. Dallai), M. *Phataginus tricuspi* (L. Liaw), N. *Uromys caudimaculatus* (W. Breed)

With so much on the line, it's no wonder sperm competition pushes male animals to get creative with their designs. These guys treat their sperm like high-performance race cars, fine-tuning every detail to snag the ultimate trophy – the egg. It's as if each sperm comes custom-built for victory, ready to out-swim and outmanoeuvre the competition in a no-holds-barred race to the finish line (Figure 6.3).

Morphology reflects function and action, and sperm are no exception – they really do compete, often in ways beyond the stretch of our imagination. Take deer mice from the genus *Peromyscus*, for example. In promiscuous species, where the race to fertilize eggs is fierce, a male's sperm often team up, forming tiny squadrons that swim faster together than solo. This synchronized swimming is no accident; it's powered by the gene *Prkar1a*, a metabolic maestro that acts like a throttle, revving up or slowing down the sperm's tail-whipping propulsion.[8]

But *Prkar1a* isn't flying solo. So far, forty-five genes have been identified that play a similar role across species like fruit flies, ants, roundworms and rodents.[9] This adds a whole new layer of complexity to the hidden world of sperm competition, showing just how sophisticated this tiny, but crucial, battle really is. It also gives us some new perspectives about the intricacy and intrigue in this less-known haploid form of existence.

For the male, the most critical mission is giving his sperm the best shot at success, so maybe it's no surprise that this is where the real drama unfolds. For instance, in the black-winged damselfly the male uses his penis like a mop, sweeping out any leftover sperm from previous suitors before depositing his own.[10] Similarly, in birds like the European dunnocks males peck at the female's rear end to make her expel the sperm from earlier mates before they get down to business.[11] Male garter snakes take it even further – they leave behind a jelly-like plug after mating, a natural chastity belt, to keep other males from getting in on the action. And if that wasn't enough, they also spray the female with a scent – a disgusting repellent – that makes her less attractive to other suitors, just to seal the deal.[12]

Make no mistake, despite all the clever tricks out there, the basic strategy for winning sperm competition is still brute force: crank out more sperm and overwhelm your rival's, just like in old-school warfare where the bigger army with more soldiers usually won the day. Expectedly, the more promiscuous the females, the tougher the competition, and the more males have to invest in larger testes to keep up with the demand.[13] This leads to an all-out arms race in sperm production. In this high-stakes battle, cutting corners just isn't an option – if you don't keep up, you're out of the game.

My appreciation for sperm competition came in a rather unforgettable way. In 2001 I was invited to a research seminar at the Max Planck Institute. My host, primatologist Ulrich Reichard, treated the seminar speakers to a visit to the bonobo exhibit at Leipzig Zoo. As we watched, a male bonobo groomed himself, his scrotum boasting testicles the size of two softballs. Ulrich glanced at the impressive display and remarked with a wry smile, 'Some people can't tell bonobos from chimps. Is it really that hard, Lixing?'*

Beyond just size, males have another trick up their sleeve: they may ramp up sperm production and increase mating frequency when they sense their paternity might be at risk. For example, when male meadow voles catch a whiff of another male's scent, they kick their sperm production into overdrive during mating, making sure they've got the best shot at fatherhood.[14]

There is, of course, a catch to all this. Since energy and material resources are limited, you can't pour everything into making sperm, and this is where the two forms of existence come into conflict for a male Dawkins's vehicle. You have to

---

* Bonobos are more promiscuous than chimpanzees. They often use sex to build social bonds and ease tension. Because of this, males face more sperm competition. To keep up, bonobo males evolved larger testicles to produce more sperm and improve their chances of fathering offspring.

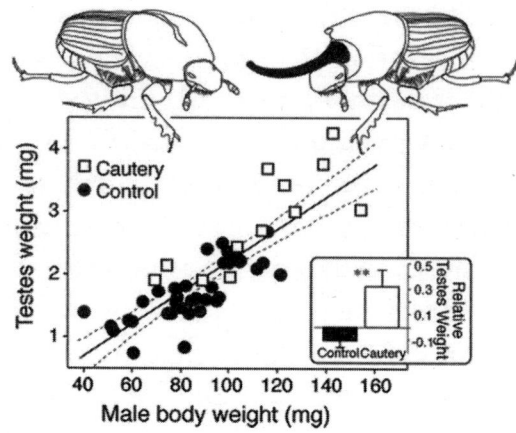

Figure 6.4. Males dehorned at the third larval instar stage (left/cautery, labelled as open squares) invest more in testes than horned males (right/control, labelled as dark dots)

strike a balance between sperm production and other vital functions, and this trade-off can lead to some creative reproductive strategies in animals.

Horned scarab beetles illustrate this perfectly. Dominant males grow impressive horns (Figure 6.4), but the cost is steep – so much energy and resources go into horn growth that other body parts, including the testes, pay the price. Meanwhile, smaller, less well-fed males can't afford the luxury of big horns. Instead, they funnel their limited resources into larger testes. While they may lose in head-to-head combat, they've got a different strategy: winning where it counts in sperm competition.

The same principle applies to marine mammals like elephant seals and walruses. Dominant males invest heavily in weapons like tusks and large body size (including big genitals), but as a result they often have relatively smaller testes,[15] creating an opportunity for smaller males to win in the sperm competition by strategically investing more material and energy in sperm with more quantity or better quality. Interestingly, this trade-off seems to apply to humans as well. In a study of 118 Caucasian men, Australian researcher Yong Zhi Foo and his team uncovered an ironic twist: the macho men who turn heads with their brawny physiques often have lower-quality sperm.[16] It's a reminder that in the game of reproduction, sometimes the underdog has the last laugh.

Lastly, males face another hurdle in the race-of-sperm competition: many female animals have a special storage system in their bodies where they keep sperm before using them to fertilize their eggs, which are produced and released over a stretch of time. Typically, it's a 'last in, first out' situation – just like how the last tennis balls you put in a container are the first ones you grab when you practise. This means that the mating sequence is crucial, generally favouring the last male to mate with the female.

Thus another set of tactics emerges, where males aim to take advantage of this last-male benefit. There are two main strategies. First, mate often to increase the chances of being the last one in. Second, prevent other males from getting their chance, usually through mate guarding – a tactic we'll dive into later in the chapter.

• • •

According to Bateman's rule, the sex with higher reproductive potential usually ends up competing more fiercely for mating opportunities – typically, this means males. But here's the catch: mate competition is mutually destructive, limiting the reproductive success of most males. Since all but a fraction of male reproductive capacity goes to waste, females – despite producing fewer eggs – may actually have greater fitness potential in some ecological conditions for species such as jacanas and clownfishes. When this happens the tables can turn, leading to females competing for males and the resources they control.

It's no surprise that direct female competition is common in species with gender reversal, like *Neotrogla*, jacanas and spotted hyenas. In these species, females often evolve larger body sizes and special weapons to fight for males, securing better mates, more food, protection or paternal care.

Even in species without complete gender reversal, female competition for males can be fierce. In Mormon crickets, males produce a giant spermatophore – a nuptial gift that can weigh 25–30 per cent of their body. Naturally, females fight tooth and nail for these males, whereas the males become choosy, preferring larger females.[17] In cardinalfish the situation is even more interesting: it's the males who carry the babies – in their mouths. While babysitting, they can't eat or flirt, and their mouths can only hold so many eggs. That makes them the bottleneck. So instead of males chasing females, it's the females who hustle – competing to stuff their eggs into as many male mouths as possible before the space runs out.[18]

Then there's the triplefin blenny – a cave-hugging fish with a flair for drama. In this species, it's the females who compete for access to nesting males. When one manages to snag a mate, others sometimes lurk nearby, watching and waiting. The moment the male releases his sperm, these sneaky sisters dart in and lay their own eggs, hijacking the fertilization in a blink.[19] This mirrors a strategy used by males in several species of salmon: precocials mature quickly and sneak mating opportunities.

Just like males, females can be downright ruthless when it comes to mate competition. They'll pull out all the stops, even if it means sabotaging their rivals' chances just to boost their own. Female antbirds (*Hypocnemis peruviana*), for example, will belt out jamming songs to disrupt a competitor's mating opportunities.[20] Female mice are even more cunning: they secretly release a pheromone in their urine that messes with the hormones of nearby females, putting their reproduction on hold.[21] Meerkats take it a step further in spiteful behaviour: dominant pregnant females will kick out gestating subordinates from the colony, imposing so much stress that it can cause them to abort.[22]

Female competition among our primate relatives can be as fierce and nasty as in any other animal group. Monkeys like woolly monkeys and Tibetan macaques show a particular mean streak – females will harass and disrupt others in the middle of mating.[23] In a gorilla society, where the dominant silverback hogs all the mating rights, dominant females will mate with him repeatedly, even while pregnant, likely to drain his sperm supply and keep other females from conceiving.[24] Chimpanzee females vie for higher social ranks, which give them better access to food and other resources, and their rivalry can be brutal, with dominant females displaying aggressive behaviour towards new female immigrants.[25]

This isn't just about mating either; some females take their rivalry to a darker place. In many mammals – rodents, rabbits, carnivores and primates (including chimpanzees) – females can be fiercely protective of their own offspring while turning into Cinderella's wicked stepmother around others' babies.[26] Aggression towards, or even killing, another female's young, while evolutionarily necessary for kin discrimination, is a grim reminder that not all in nature is warm, cosy and uplifting.

・・・

When females mate with other males, mate guarding is a no-nonsense strategy for males to make sure the kids are theirs. In the animal kingdom, mate guarding comes in a variety of flavours. The most common tactics include sticking close to the female, chasing away rivals who dare to come near, and mating with the same female repeatedly to win the game of sperm competition.

Some lesser-known strategies for mate guarding have also evolved. For instance, certain mammals, like garter snakes mentioned earlier, produce a gelatinous plug that blocks the female's genital opening, effectively sealing the deal and preventing her from mating with other males. In other cases, deception is the name of the game. Males in species like barn swallows and Formosan squirrels will produce fake alarm calls right after mating. The ruse sends any nearby rivals running for cover, leaving the male with a better chance of securing his paternity.[27]

Unsurprisingly, humans have come up with all sorts of ways to guard mates – chastity belts, foot-binding, head and body covering, you name it – directly or indirectly controlling women's sexuality. But despite our creativity, we might not be the most dedicated mate guarders. That honour probably goes to the yellow garden spider. After mating, the male spider sticks his sperm-delivering 'limbs' into the female's genital opening.* While she munches on him for nutrition, his limbs stay put, acting as a natural chastity belt.[28]

As the saying goes, there's no such thing as a free lunch, and this couldn't be truer for mate guarding, which demands

---

* The 'limbs' here are not legs. They are specialized reproductive appendages called pedipalps.

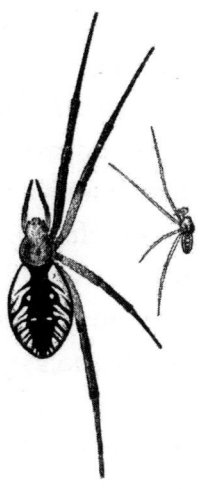

Figure 6.5. A female (left) and a male (right) garden spider (shown only half of each)

time, energy and a willingness to risk life and limb in brawls with rivals and run-ins with predators. For these reasons, males don't exactly love the mate-guarding grind, as shown in a cheeky study of Seychelles warblers. These island songbirds usually lay one egg per clutch – maybe two if the beachfront real estate is particularly nice. During the fertile period the males are all business, sticking close to their partners to fend off rivals. But here's the twist: slip a fake egg into the nest and the males are duped. They take it as a sign that their job is done, pat themselves on the back, and abandon their posts like it's 5 p.m. on a Friday.[29]

Mate guarding is not a male monopoly, either; females may also do so, especially when their mates attempt to pick up other females. One interesting scenario comes from the burying beetle (*Nicrophorus defodiens*), a species that lay eggs in the carcasses of mice, voles, shrews and others. When a couple find a dead animal, the male – being a male – may creep to an elevated area to 'call in' another female by emitting a sex pheromone. But if his mate senses it, she will rush to the spot and push the male off the ledge, as if screaming, 'Get off here, you pig!' She has no intention of having her offspring share the vital food with another female's.[30]

Birds like wrens, boubous and great tits have their own ways to keep their male partners in check, much like burying beetles. Female birds often chirp duets with their mates, not because they love a good singalong but to stop the males from belting out solo tunes that might attract other females.[31] So it's mate guarding by music – turning a romantic duet into a clever strategy to keep the competition at bay.

. . .

In humans, the relatively mild degree of sexual dimorphism points to a modest level of sexual selection in our evolutionary past. On average, men are about 25 per cent heavier, 7.6 per cent taller, and as much as 64 per cent stronger in muscle strength than women[32] – noticeable, but tame compared to gorillas, where males can be twice the size of females, or chimpanzees, where sperm competition is fierce. Researchers have linked this drop in macho intensity to a suite of uniquely human traits: concealed ovulation, pair-bonding, shared parenting, and the lack of large canines.[33] The exact evolutionary chain of cause and effect is still murky, but one thing's clear: compared to our primate cousins, human males are surprisingly civil. Macho-lite, if you will.

Even with reduced mate competition, human males still sport an impressive line-up of secondary sexual traits – broad shoulders, big muscles, deep voices, chiselled jaws and, of course, beards. These features have evolved less to charm women than to intimidate rivals. In fact, many of them are signals of dominance rather than sex appeal.[34] Male–male competition also helps explain why men have larger hearts, lungs and more red blood cells than women – traits that support physical exertion and endurance, whether for fighting, hunting or showing off.[35] And the pay-off is clear: in non-industrial populations, men with higher social status tend to enjoy higher reproductive success.[36]

Today, the echoes of our evolutionary past can still be heard in our instincts. Traits like male attractiveness, intelligence, ambition and social status are often seen by other men as competitive qualities. Similarly, men who flaunt their wealth with flashy cars and other luxury items are more likely to be viewed as rivals than friends.[37]

What makes humans truly sexually dimorphic might have more to do with what's 'between the ears than between the legs', though. Thanks to our large brains, men excelled in creating and using handheld weapons in organized conflict, long before modern technologies emerged.[38] This history may also explain why men around the world are far more interested and involved in competitive sports – especially team sports like football, basketball, baseball and cricket – than women.[39] And, of course, wars and armed conflicts have always been overwhelmingly a male affair, even though women, especially in democratic nations, have increasingly joined the ranks today.

Despite general trends, the intensity of mate competition varies greatly among societies and over time. The key factor is reproductive variance, meaning the degree to which some men have many more offspring than others given a set of social and ecological conditions. In stateless societies, men's reproductive variances can be one to three times higher than those of women.[40] In tribal societies, mate competition often involves violence, including homicide, feuding and warfare[41] – a large-scale version of a Pitcairn community.

Agricultural civilizations not only led to larger societies but also concentrated resources and power at the top, resulting in exaggerated reproductive variation among men, often more extreme than in animals with large harems, like gorillas and elephant seals.[42] This fitness skew in favour of high achievers – emperors, kings, chiefs – created a major reproductive incentive for large-scale aggression and territorial expansion.

With so much at stake in reproductive success, men, especially young men, have a higher propensity to take risks (we often adopt a cavalier attitude – 'let boys be boys' to explain this away). This is why far more men than women engage in drinking, gambling, drug abuse and other dangerous behaviours, leading to higher mortality rates among men. As we can expect, male violence remains a persistent and challenging social problem.

Though homicide rates have significantly declined, the pattern of male-dominated violence persists, much like what was observed in the mutineer community on Pitcairn. Statistics show that 95 per cent of same-sex killings are committed by men against other men, with violence peaking in the eighteen to thirty age group.[43] This isn't a phenomenon unique to industrialized societies in the West, contrary to what Rousseau believed. In some modern tribal societies, 'killing men and capturing women' remains a discernible pattern. For example, studies of the Yanomamö and other traditional societies in South America reveal that about 30 per cent of men die due to violence.[44]

Encouragingly for us humans, while Darwin's ghost still lingers, it's far less deadly now. We – especially men – are far less likely to be killed today than at any other time in history.[45]

In contrast to men, same-sex homicide is far less common among women; physical violence simply isn't as advantageous for women as it is for men. The most obvious biological reason is the vast difference in reproductive potential, leading to a sharp strategic contrast between males and females in their evolutionary quest for fitness, as predicted by Bateman's rule. Unlike men, women's reproductive success is less of a high-stakes sprint and more of a steady marathon, unfolding gradually over their fertile years and beyond. There's no short-term reproductive glory that justifies taking high risks for immediate rewards. In other words, women's reproductive strategy is long-term, accumulating fitness gradually – much like value investing, making cautious decisions and patiently waiting for long-term gains.

That said, women do compete for resources and, at times, for desirable men. Women, similar to their counterparts, frequently see other women, rather than men, as their main rivals. They can become highly critical when they see other women wearing provocative clothes,[46] and far more women than men hold negative views of promiscuous peers.[47] In some situations, women may compete directly for men. For instance, in both Eastern and Western societies where monogamy is the norm, many women aim to marry men who have resources.[48]

When it comes to competing for attractive men, women tend to be more indirect. Psychologist Joyce Benenson outlines what I paraphrase as three principles that govern competition among women.[49] First, 'do no harm'. This principle arises from the evolutionary need to avoid injuries that could hinder their long-term reproductive success. Second, 'enforce fairness'. Women strive to prevent a few high achievers from monopolizing resources, which also fosters trust and cooperation within the group. Finally, 'avoid direct competition'. This often involves excluding certain women through tactics like bullying or spreading rumours, thereby minimizing rivalry.

A typical form of aggression among women is popularly characterized by the 'mean-girl' strategy, which includes behaviours like shaming, gossiping, rumouring and ostracizing rivals. These tactics are used to boost their own popularity and gain access to desirable mates.[50] Research shows that when an adolescent girl is perceived as attractive, her chances of being targeted by indirect aggression from hostile peers can increase by as much as 35 per cent.[51]

While women can engage in physical competition for mates, the intensity and frequency of violence are far lower compared to men. This suggests that Hobbes's idea of a 'state of nature' aligns more with male behaviour than female behaviour. One reason for this difference? For women, cooperation matters just as much as competition. A strong social network isn't just a bonus – it's a key ingredient for reproductive success. The same holds true for our primate cousins like macaques and baboons, where tight-knit female alliances offer a serious advantage – babysitting, food-sharing, mutual aids. In these societies, social bonds aren't just about friendship; they're a survival strategy, especially for mothers raising young.[52] This is why female relationships often involve both love and rivalry for us and for many of our primate relatives. It is no coincidence that even on Pitcairn Island, nine of the last ten survivors of the Bounty were women.

Red in Tooth and Claw

We began and ended this chapter with the species we know best – ourselves. But the real marvels of mate competition come alive in other creatures. Evolution struts its stuff in thunderous songs and flamboyant dances, in antlers locked in battle and feathers flared in show. All this theatre, all this fire, owes its spark to sex. Yet behind the curtain another force of sexual selection is stirring. Not brute strength, but choice; not battle, but desire. So let's turn the page. A new chapter awaits – where beauty wins the day, and preference becomes power.

7

A TASTE FOR THE BEAUTIFUL
Mate Choice

Beauty pageants are an irresistible cultural spectacle. Every year, millions tune in, eyes glued to dazzling live broadcasts like *Miss Universe* and *Miss America*. Contestants embody picture-perfect ideals – carefully screened so they're never too short, too tall, too fat, too skinny, or too plain. Even in pageants claiming to value talent, intelligence or charity work, these virtues tend to fade into the background, overshadowed by the glittering parade of swimsuits, stilettos and sequinned evening gowns.

The so-called talents on display are usually nothing more than cheerful fluff – singing, dancing and instrument-playing are practically mandatory. You'll almost never see a contestant flaunt her skills in maths, science or business savvy. Can you imagine someone whipping out a blackboard, mid-pageant, to dissect the Red Queen hypothesis or unravel the Schrödinger equation?

Regardless of your opinion on beauty pageants – and they can admittedly be astoundingly frivolous affairs – Western beauty pageants occasionally serve up moments of spectacular gibberish, but at least they've made real strides towards inclusivity, welcoming contestants from different ethnic backgrounds. The same can't always be said in parts of Asia.

A Taste for the Beautiful

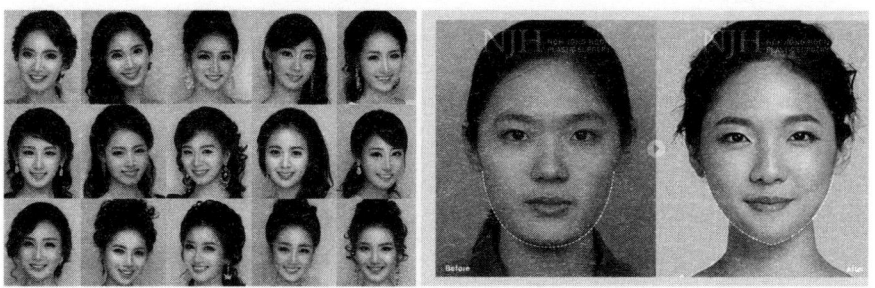

Figure 7.1. Left: Finalists for 2013 Miss Daegu in South Korea. Right: The transformation of an ordinary person to a glamorous beauty before and after cosmetic surgery[1]

In South Korea, for instance, pageants still showcase contestants so eerily similar in body shape, facial features, fair skin and large eyes you'd swear they were all sisters posing for a family portrait (Figure 7.1).

Why is this the case, you may ask? The answer: they are custom-made. This is only somewhat tongue-in-cheek. Beyond heavy make-up, many contestants for Miss Daegu (a South Korean beauty pageant held in the country's fourth-largest city) and other pageants undergo extensive cosmetic surgery to enhance their looks: body shape, facial features, even height. Getting plastic surgery in South Korea can cost more than what the average person makes in a whole year – and good luck getting your insurance to chip in. But that hasn't stopped anyone. Fuelled by sky-high demand and juicy profit margins, South Korea has become the undisputed plastic surgery capital of the world. Despite the price tag, one in five Korean women have gone under the knife. In Seoul, it's closer to one in two.[2] And it's not just locals. Tourists from China, Japan and beyond pour in, hoping to tweak, smooth, lift or reshape whatever part of themselves they've been taught to call a 'flaw'.

Behind these painstaking and often painful quests for physical perfection lies a collective societal mindset about what constitutes beauty. Any deviation from these rigid standards is deemed unattractive and often met with social disapproval. If something about this feels off to you, you might find yourself pondering a deeper question: where did our aesthetic sense – the ability to appreciate beauty – come from in the first place? The short answer is mate choice. This will be our focus for the chapter, and we begin with one of the most striking examples in the natural world – the peacock.

. . .

'The sight of a feather in a peacock's tail,' Darwin confided to Harvard botanist Asa Gray in an 1860 letter, 'whenever I gaze at it, makes me sick.' This 'sickness' wasn't revulsion. It stemmed from a deep unease due to a perceived contradiction with his theory of natural selection. That flashy fan of feathers didn't just slow the bird down; it practically screamed 'eat me' to any nearby tiger, leopard and other predators. It was a glittering death trap. How, Darwin wondered, could evolution favour something as outrageous – and impractical – as a peacock's tail? Eventually, he offered a bold hypothesis: maybe it wasn't about survival at all. Maybe the peahen simply had a 'taste for the beautiful'.

In his 1871 book *The Descent of Man*, Darwin introduced the concept of mate choice as a force of sexual selection, distinct from the brutal competition for mates. In one form of sexual selection, he wrote, individuals – usually males – compete with each other, sometimes violently, to defeat or drive off their rivals, while females stay passive. In the other form, males still compete, but instead of fighting they try to attract or impress members of the opposite sex – usually females – who are no longer passive observers but actively choose the partner they find most appealing.

However, whether peahens truly have a taste for the beautiful remained unanswered until 1987, when Darwin's modern compatriot Marion Petrie decided to test the idea in a free-ranging population of about 180 individuals at Whipsnade Park, 35 miles north-west of London. As we now know, sexual dimorphism – especially in size difference between males and females such as in elks, narwhals and gorillas – often signals mate competition. In the peacock, however, its ungainly tail is an obvious burden to flight, and far from a useful weapon in a fight. Could this dazzling display be the result of female preference? And if so, what exactly might appeal to the discerning peahen?

Marion focused on the most striking feature of the peacock's tail: those shimmering eyespots scattered across the feathers like jewels. In the name of science, she snipped off a few and watched what happened next. The results were brutally clear. Males with fewer eyespots endured a sad and lonely mating season, while their untouched rivals strutted around, winning peahen hearts left and right.[3] Marion's verdict? Peahens definitely have an eye for beauty – and beauty, it seems, is all about counting eyespots.

Marion's findings caught the attention of biologist Mariko Takahashi – but Mariko wasn't fully convinced. Determined to double-check these findings, she gathered a team in Japan to replicate the experiment. Seven long years later, Mariko had her own story to tell. Her peahens, surprisingly indifferent, shrugged off the eyespots altogether. Apparently, counting eyespots wasn't their idea of romance after all.[4] In North America, meanwhile, researchers Roslyn Dakin and Robert Montgomerie also found that most peacocks had a similar number of eyespots, leaving peahens with little to choose from.[5] It seems that clipping eyespots, as Marion had done, was more a side effect of human meddling than something that happens in nature.

Marion felt compelled to go back to the drawing board. By 2009 – twenty-two years into the project – she and her team finally cracked the code of peahen desire. Turns out, the critical 'X Factor' was the length of the peacock's train – the feathers which cover the base of the tail. Their results got a high-tech stamp of approval when Jessica Yorzinski's team used an eye-tracking device to confirm exactly where peahens were looking.[6] Marion may never have expected that testing Darwin's idea of 'an eye for the beautiful' would take her nearly an entire academic career before the record could be set straight.

Hold on a second, you might be thinking. If peahens only care about the train, then what's the deal with those dazzling, iridescent eyespots? Are they just for decoration? Turns out,

Figure 7.2. A blue peacock

they do have an important job. Peacocks grab the attention of far-off peahens by calling, rattling their trains, shaking their wings and shimmering those flashy eyespots. But once the ladies wander in for a closer look, the magic fades – the eyespots don't seem to matter much any more.[7] So, in the world of peacocks, beauty – a trait attractive to mates – depends on distance.

Why do peahens have an eye for the beautiful, despite the influence of distance? To understand this we need to revisit the distinction between the two forms of existence: the diploid body and the haploid gametes (sperm and eggs). In most animals, the diploid body does all the hard work, looking after those carefree, haploid gametes. During mating the body plays matchmaker and chauffeur for its gametes, carefully screening potential partners to secure the brightest possible future for its genetic offspring. This choosiness matters a whole lot when gametes are limited or costly to produce – think of those precious, hard-earned eggs or, under special conditions, even sperm. That's exactly where the drama of mate choice kicks into high gear. And somehow, I found myself caught right in the middle of it.

I've always been captivated by the mysteries of mate choice: why do animals choose the mates they do? In 1996 I found myself headed to the perfect place to dive deeper: Mike Ryan's lab at the University of Texas at Austin. The very same place where Hermann Joseph Muller, of Muller's ratchet, made waves some seventy years earlier.

On a chilly January day, I began my journey to the Lone Star State, opting for an Amtrak train over a plane just to save $40. It might not sound like much, but back then it was enough to keep my wife and me fed for two whole weeks.

The train was delayed for hours, and by the time I rolled into Austin it was well past midnight. I figured that, in this pre-cellphone age, there was no way Mike would know and wait to pick me up and had already resigned myself to spending the night on a bench at the station. But lo and behold, there he was – Mike, with his little Mazda RX-7, bald head shining in the streetlight and a short black beard that would make a middle-aged Charles Darwin do a double-take.

Born in the Bronx to a truck driver, Mike was the eldest of eleven kids. In a time when boys were expected to follow in their father's footsteps, Mike's interests took a wild detour. Trucks? Not a chance. Mike was all about nature – whether it was tiny grasshoppers or towering dinosaurs. His parents quickly learned that the best way to make him happy was a trip to the Bronx Zoo or the American Museum of Natural History.[8]

After high school, Mike headed to Glassboro State College, set on becoming a high school biology teacher. But everything changed after a trip to the Galapagos Islands. 'It was my first time leaving the country, my first time on a plane, and my first time in the tropics,' he recalled. That trip hit him like a lightning bolt. Inspired by Darwin's adventure, Mike scrapped his teaching plans – he was 'seduced by research' and went to Rutgers University to get a masters degree in biology.

While studying bullfrogs in 1976 and 1977, a nagging question got under his skin: why do frogs make those loud, ridiculous mating calls? That curiosity drove him to pursue a Ph.D. at Cornell University, where he soaked up cutting-edge ideas and technologies like Fourier transformation and information theory, all the while rubbing elbows with star scientists like Carl Sagan and Jane Goodall.

Fuelled by a burning desire to uncover the truth behind mating calls and mate choice, Mike set off for the jungles of

Panama. His target? The red-eyed tree frog, that bright-green species you've probably seen gracing a million posters. But his well-laid plans unravelled fast. Not only did he face a barrage of sticks thrown by unwelcoming capuchin monkeys up high in the canopy, he also ran into a more fundamental problem: evolution had left him woefully unequipped for life in the treetops. Try as he might, Mike often couldn't climb high enough to record mating calls from those elusive frogs.

But there was another reason that still makes him laugh today: 'the din of those bothersome chorusing túngara frogs at my feet made it nearly impossible to record the mating calls of the treefrogs'. These small, brownish frogs might not be poster material but they were everywhere, turning puddles and swamps into noisy, froggy raves. They were practically begging, it seemed, to be studied. And that's when it hit him. 'Why not?' he murmured to himself. And that moment, he struck scientific gold.

Still, recording frog calls in a tropical rainforest was no walk in the park, especially with the clunky technology of the time. For the next three years Mike lugged around a 17-pound Nagra tape recorder, shielding it from the ruthless rain and clutching it like a lifeline whenever he slid down muddy hillsides. Thanks to endless patience, stubbornness and a growing pile of destroyed boots, he managed to identify 617 male túngara frogs, document 751 mating events, and run a full roster of experiments – Mike was now fluent in the frog's flirting language.

Darwin hypothesized that females might have a taste for the beautiful. At the same time, he admitted, 'it is, however, difficult to directly obtain evidence of their capacity to appreciate beauty'. Mike was the first to change that statement to the past tense. He showed that sexual selection by mate choice wasn't just a theory – it was happening right there in the rainforest, one froggy croak at a time.

Mike discovered that male frogs have a hidden trick up their sleeves – or more accurately, down their throats. Their mating call isn't just one sound, but two: a constant, high-pitched whine that's unique to their species, plus one or more optional low-pitched chucks. Female frogs, it turns out, have a soft spot for deeper calls from the bigger males. But here's the kicker: females really lose it for males who can mix in a bunch of chucks, which scream 'I'm sexy!'.

Why, you may wonder, aren't all the males belting out their sexiest chucks like they're auditioning for *Frog Idol*? The answer, as ever, lies in the trade-off of these chucks, which attract the attention of frog-eating bats, making this high-reward mating call a risky move. Apparently, only the male frogs slick enough to dodge their winged assassins – the bats – can manage to croon this flirty song. So could a taste for beauty actually be a shortcut to better genes? Mike didn't have the numbers to back that up. But he found an answer in Marion's peafowls: peahens who fall for peacocks with the longest, flashiest tails end up with chicks that are healthier and tougher with a better chance of survival.[9]

It's easy to see why females fall for males offering clear perks – prime territory, sturdy protection, devoted parenting or nuptial gifts.[10] Take the fifteen-spined stickleback, for example: females swoon over males who excel at fatherhood – meticulously building nests and handling all the childcare duties. In this species, a male's beauty lies in how well he can shake and shimmy, vigorously fanning fresh, oxygen-rich water over the eggs with his body.[11] But frogs and peacocks? They're not exactly winning 'Father of the Year' awards.

They mate and vanish, leaving females with no obvious benefits beyond good looks or sweet serenades. So why be choosy at all?

The answer lies in something called indirect benefits – the evolutionary version of shopping for 'good genes'. In theory, by picking flashy mates females could pass on superior DNA to their offspring, giving them a better shot at surviving and thriving. So far, so good. Except that evolutionary biologists Angela Achorn and Gil Rosenthal recently poured a bucket of cold water on that cosy idea. After digging through piles of scientific studies, they found surprisingly little evidence to support the classic 'beauty equals good genes' story.[12] So the jury is still out.

Even so, beauty exists in many forms in the animal world. One popular form of beauty is bright colours. In the three-spined stickleback, for example, male beauty is all about that striking red belly – a sign of top-notch health and fewer parasites. That red colour comes from a diet loaded with carotenoids, which are powerful antioxidants. But it's not just about looking good; a redder belly also means a better dad, one who can fan more water into the nest to help the eggs develop.[13]

Carotenoid-related bright colours – orange, yellow, red – are a common beauty standard in birds too. From ibises and flamingos to songbirds like finches, canaries and cardinals, it's all about showing off those vibrant hues.[14] In northern cardinals, for instance, females go wild for males with the reddest feathers because those guys are better providers, bringing more food to their chicks than their duller counterparts.[15] Apparently, this carotenoid-based system is a tried-and-true method for females to pick mates who offer both direct benefits (like good food and attentive parenting) and indirect benefits (like the putative 'good genes' for offspring survival).

On the Origin of Sex

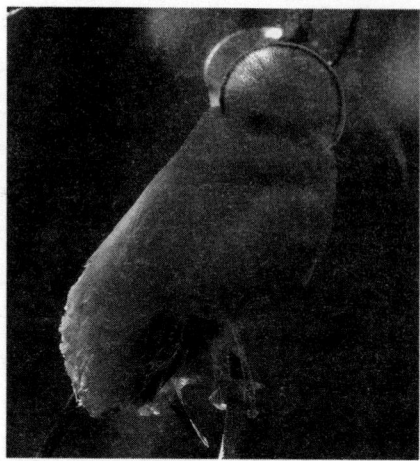

Figure 7.3. A male Guianan cock-of-the-rock, one of my favourite birds, demonstrates a high level of carotenoids through its bright plumage

But beauty in birds isn't just skin-deep – or feather-deep, for that matter. In many songbirds, beauty is also about a male's ability to sing his heart out. In canaries, a male's allure comes down to how intricate and mesmerizing his tunes are. For song sparrows, it's less about the flair and more about the playlist – males with the most songs win the spotlight. Then there are golden-collared manakins, where beauty lies in endurance – males win over females by performing long, energy-draining courtship dances bolstered by high metabolic rates.[16]

But it's not all about songs, looks or dance moves.* In birds like crows, ravens and parrots, beauty is in the brains. These clever birds need serious cognitive skills to thrive in their complex environments, and as you might guess, females

---

* In a classical study by William Hamilton and Marlene Zuk, ornamental plumage in birds was shown to signal an individual's ability to withstand parasitic infections. (Hamilton, W.D. and Zuk, M., 1982. Heritable true fitness and bright birds: a role for parasites? *Science*, 218(4570), pp. 384–7.)

in these species go for the brainier males. Even parakeets get in on the action, with females choosing males who can solve puzzles related to food extraction – clear proof that a sharp mind is just as attractive as a pretty face or feathers.[17]

In mammals, beauty often means diversity when it comes to the major histocompatibility complex (MHC) system. The MHC is like a biological security team, made up of dozens of genes that are pros at spotting pathogens. If a female picks a mate with different MHC genes, she's giving her kids an upgraded defence system with weapons she herself doesn't have. This means her offspring will be better equipped to detect and fight off a wider range of diseases. And many mammals, especially rodents, do choose mates whose MHC genes are different.[18] Do humans do the same? Apparently, yes.* In some studies, women show their abilities to prefer odours from men whose MHC genes differ from their own.[19]

Make no mistake, mate choice is no passive affair – females are actively calling the shots. When they're on the fence about which male to choose, they often watch and learn from their peers, a behaviour known as mate choice copying. You'll see this in fish, birds and even mammals – including humans.[20]

But the barnyard chicken takes active mate choice to a whole new level. Hens have an extraordinary ability to eject most of the sperm they receive after mating with a low-

---

* A recent meta-analysis of existing studies indicates that women's preferences for men with dissimilar MHC profiles are not statistically significant, suggesting that additional, as yet unidentified, factors may influence the outcomes of such studies. (Havlíček, J., Winternitz, J. and Roberts, S.C., 2020. Major histocompatibility complex-associated odour preferences and human mate choice: near and far horizons. *Philosophical Transactions of the Royal Society B*, 375(1800), p. 20190260.)

ranking male,²¹ as if to say, 'Oops, that was a mistake!' This clever method works a lot like our own Plan B, the Morning-After Pill – except these hens are doing it with flair, no pharmacy required.

The hen's example also shows that female mate choice can be hidden, sometimes playing out inside the female's body, especially in the reproductive tract or eggs.²² As we saw in the previous chapter on sperm competition, fertilization often follows the physical rule of 'Last in, first out'. But in many animals, females have the power to flip the rule on its head, though exactly how they pull off this magic trick is still a bit of a mystery.

Some female mammals have a secret weapon – they can selectively allow sperm from preferred mates to fertilize their eggs after multiple matings. This cryptic choice gives them a powerful way to ensure that only the sperm from their preferred males gets the green light. Bats, in particular, take this to the next level: in some species, the implantation of fertilized eggs can be delayed for weeks or even up to a year, giving the female time to perform a kind of genetic screening. During this waiting period, she can discard some – presumably low-quality – embryos, letting only the best candidates implant and develop further.²³

So far, we've been talking about females as the choosy ones, but male mate choice is also common in the animal world. Sure, sperm might be cheap and plentiful, but they can't be sent out one at a time. Instead they're delivered in batches as ejaculates, each packed with tens or even hundreds of millions of sperm cells. And in some species – insects, amphibians and others – these ejaculates come with a little something extra – nuptial gifts – to sweeten the deal in the mating market. Given the energy and resource cost of this, a certain level of choosiness on the part of males is prudent.

Ejaculates, unlike the individual sperm cells they contain, are limited, much like eggs. This puts a dent in Bateman's rule, which says that males are eager and undiscriminating. Even in fruit flies, where males don't provide any paternal care, they still prefer to mate with more fecund females.[24] In topi antelopes, where females compete fiercely for mates, attractive males often face sperm depletion. To make the most of their limited resources, then, these males prefer to mate with unmated females, ensuring their precious ejaculates aren't wasted.[25]

Sometimes male mate choice can be stronger than female mate choice, especially when males are in short supply or when gender roles are reversed. This happens in certain fishes like seahorses, pipefishes and scissortail sergeants,[26] as well as in birds and mammals like jacanas and spotted hyenas. In these species, females compete for mates while males pick females of their preference, often those with larger bodies. In such cases females too can flaunt their beauty to attract males.

Female dance flies, for example, puff up their bellies to look hyper-fertile, luring males who arrive bearing lavish nuptial gifts.[27] Similarly, female two-spotted goby fish flash their bright-orange bellies, signalling peak health and fertility to choosy males.[28] And here's a juicy twist: just like domestic hens eject sperm from second-rate mates, 'pregnant' male black-striped pipefish – a quirky cousin of seahorses – can spontaneously abort embryos they're carrying if a more attractive female swims by.[29] The Portuguese researchers who discovered this, led by Mario Cunha, cheerfully dubbed it the 'Woman in Red' effect, paying playful homage to Gene Wilder's iconic comedy. Who knew fish had their own scandalous romantic dramas?

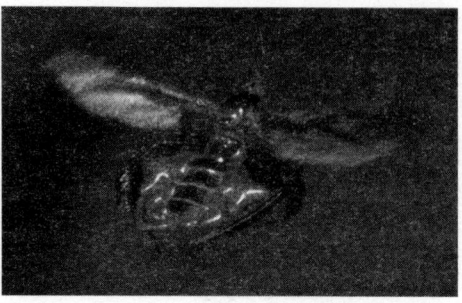

Figure 7.4. A female long-tailed dance fly flaunting an inflated abdomen to attract males

While we've only touched on a few examples, we can already see a clear pattern emerging: if mate competition is about building a sturdy machine – Dawkins's vehicle – for strength and endurance, mate choice is often about crafting a sharp mind – a cognitive ability – for wit and charm. That's why genes and neurons often steal the spotlight when it comes to picking a partner. For example, a female frog's mid-brain neurons can light up like fireworks in response to certain notes in a male's call. Or consider insects: if you swap out a receptor gene in a fruit fly's antenna with one from a moth – like getting your dating preferences rewired – the fly suddenly finds itself irresistibly drawn to moth pheromones instead of its own kind.[30]

• • •

If all the examples above demonstrate anything, it's that mate choice can have a huge impact on shaping sexual dimorphism. So the big question is: how does this process get started? Which takes us to Darwin's favourite mammal – the mandrill. In fact, 'no other member in the whole class of mammals is coloured in so extraordinary a manner as the adult male mandrills,' he wrote in *The Descent of Man*. And he wasn't exaggerating.

Socially dominant males of this species of baboon sport strikingly bright reds and blues on their faces and rear ends.

Why do male mandrills sport such wild, flashy colours – and in such odd places? To crack this mystery, we turn to an idea called the sensory exploitation hypothesis, cooked up by none other than Mike Ryan. Mike's big insight is that females set the trend first, evolving preferences before the males caught on. Only later did males evolve clever ways to tap into these pre-existing female tastes. In the túngara frog, for example, males tweak their mating calls to match the pitch that makes female frogs sit up and pay attention. They're basically broadcasting on the exact station females were already tuned into.

But in the 1990s, Mike encountered a potential snag in his theory: in the two species studied extensively in his lab – the túngara frog and the cricket frog – the males' call pitches are actually a bit higher than the sweet spot in the female ear's frequency range. It's like trying to impress a female listener whose favourite male opera singer is a baritone, but by deciding to sing as a tenor. Why the mismatch?

Figure 7.5. A male mandrill (left) and male uakari (right)

This was the puzzle I set out to solve when I arrived in Austin. Mike and I decided to compare the relative performance – that is, how well the spectrum of male calls matched female perception – at short distances (1 metre) versus long distances (11.6 or 16 metres). I took mating calls recorded in the wild to test a novel idea: maybe calls with slightly higher-peak frequencies are better at reaching females from afar. The science behind this is straightforward: higher-pitched sounds fade faster than low-pitched ones, especially when there's stuff in the way – like trees, bushes or rocks. It's why, when someone drives past you blasting rock music, you mostly hear that thumping bass, while the high notes of the lyrics vanish into thin air.

And that's exactly what I found in both frog species. As male mating calls travelled further, losing strength along the way, their pitch shifted downward – landing perfectly in the sweet spot of the female's hearing.[31] By starting at slightly higher frequencies, males cleverly ensured that their calls reached distant females loud and clear, much like how peacock eyespots capture peahen attention from afar. Mike's sensory exploitation hypothesis hits the mark again!

What works for frogs and peafowl also works for primates. Primates love fruit, a rich nutritional resource desired especially by females. But in the dense green of tropical rainforests, spotting ripe fruit isn't easy. To make matters worse, tropical fruiting seasons are long and scattered. This challenge pushed our branch of primates to evolve keen colour vision, finely tuned to the hues of ripe fruit.* So when a male monkey sports bright colours on his body, he's tapping into that evolved visual sensitivity, transforming

---

* Humans are highly sensitive to bright colours like red and orange, which is why red signals 'STOP' in traffic signs.

'come get a bite of this' to 'hey, I am here'. This gives him a competitive edge in catching the attention of females – a clever move in the mating game.

For the uakari monkey in the Amazon, males flaunt vivid red heads, as if they've been painted bright red on their bald skulls. Similarly, female mandrills are drawn to males with bold colours. The blue on their faces heightens the contrast with the red and makes them pop against the green foliage, ensuring that these vibrant males don't blend into the background. It's a neon sign in the jungle, impossible to ignore.[32]

In both cases, you can see that males evolved these flashy traits to tap into females' pre-existing sensitivity to bright colours, perfectly aligning with the sensory exploitation hypothesis. Here, we've just cracked another puzzle posed by Darwin. But there's more – mate choice doesn't just shape individuals. It can drive the formation of new species, an idea that even Mike didn't see coming.

Several years after I left Austin for a job in Washington, Mike's grad student, Kathy Boul, headed to Ecuador to study Peters' dwarf frogs in the Amazon. There she stumbled upon a paradox. Despite being only 12 miles apart, two populations of these frogs had males with completely different mating calls – one simple, the other complex. And, most importantly, the females only responded to the calls from males in their own populations.

Genetic tests and behavioural assays all pointed to the same conclusion: these two populations were on the path to becoming separate species.* The females had developed

---

* There are many ways to define a species. One common definition, called the biological species concept, states that a species is a group of organisms that can breed with each other in nature and produce fertile offspring.

their own distinct tastes for what they found beautiful, and this divergence in preferences was driving the speciation process.[33] It's a fresh example of how sex can generate diversity, adding a new twist to what we've explored in earlier chapters of this book.

· · ·

Although we've uncovered some fascinating insights into the peacock's tail, many questions remain unanswered. Among them is how the peacock got its long and elaborate plumage. For nearly sixty years after Darwin suggested that peahens have a 'taste for the beautiful', no one could come up with a convincing explanation for the question.

The first serious attempt came from evolutionary biologist Ronald A. Fisher in 1930. He assumed that male traits and female preferences for those traits are genetically linked. Imagine there's a gene – or a set of genes – involved. In peacocks this gene expresses itself as a long, elaborate tail, while in peahens it shows up as a preference for that tail. This creates a positive feedback loop, where the trait and the preference for it reinforce each other, causing the trait to evolve rapidly – technically called a runaway.[34]

However, Fisher, who was also an accomplished statistician, didn't develop a formal mathematical model for his runaway hypothesis. This gap was eventually filled in the early 1980s by mathematician Russell Lande and Mike's colleague Mark Kirkpatrick at UT Austin. Lande and Kirkpatrick demonstrated that Fisher's runaway effect isn't universal; it holds up only under certain conditions.[35]

Now the real challenge: the runaway hypothesis, despite its tidy logic, doesn't quite match up with nature. The evidence biologists have so far gathered can be described as tantalizing at best. In some species, like stalk-eye flies and stickleback fish, male ornaments and female preferences appear to be linked,[36] but the hunt for the genes predicted by Fisher's hypothesis has mostly come up empty.* Could this assumption of linked genes be the very hurdle that keeps the runaway process grounded on the evolutionary runway?

Evolutionary biologist Amotz Zahavi had a different spin on the beauty game. He proposed the handicap hypothesis, a clever workaround to the tricky assumptions baked into Fisher's runaway theory. Zahavi argued that flashy ornaments – like long tails and bright colours – aren't just eye candy. They're costly, honest signals of a male's quality. The idea is straightforward: only the strongest males can afford to show off without crashing and burning. In other words, if a male (like a peacock) can survive with a giant, showy tail slowing him down and making him a walking target, he's probably got some serious genetic chops.[37]

By ditching Fisher's assumption of genetic linkage, the handicap hypothesis packs more explanatory punch and is easier to test in the real world. In the house finch, for example, brighter, redder colours in males signal a superior diet (and hence superior feeding and survival ability), making females swoon over these vibrant suitors.[38] Similarly, in the peacock that extravagant tail isn't just for show – it's an honest signal. As Marion Petrie discovered, offspring from males with longer, more elaborate tails tend to grow faster and stay healthier.[39]

It makes sense that a costly trait can weed out the fakes, but does a handicap really need to be as over-the-top as the

---

* Fisher did not assume that the trait and the preference for the trait should be physically linked, such as two genes located on the same chromosome. Nevertheless, contemporary researchers often employ linkage mapping to identify genetic associations between traits and preferences.

peacock's tail to be an honest signal? Couldn't a slightly larger tail than his rivals' do the trick of attracting mates while saving the peacock a ton of energy? Surprisingly, the answer is no, and the reason lies in a little-known principle called the Weber–Fechner law, named after two nineteenth-century German scientists, Ernst Heinrich Weber and Gustav Theodor Fechner.

Before diving into the details of the law, let's get a feel for it first. Take a look at the two pairs of pictures in Figure 7.6 and compare the density of dots. Which difference is easier to spot: the one between ten and twenty dots or the one between 110 and 120? For most people, the difference between ten and twenty is far clearer, though the quantities differ by the same amount. What's behind this?

It turns out that animal sensory systems, including ours, are wired to spot distinctions based on relative scale, not absolute amounts. This principle appears to apply across all senses within a certain range because our sensory neurons respond to stimuli based on relative intensity, not absolute

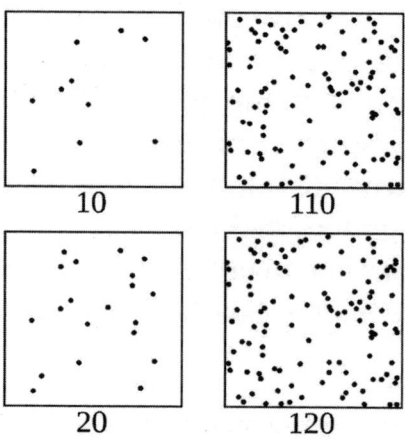

Figure 7.6. An illustration for Weber–Fechner law

values.[40] So even though the difference between 20 and 10 and between 120 and 110 is both 10, it's a lot harder for us to see the gap between 120 and 110. In other words, if 20 is twice as much as 10, we'd need 220 for the same effect compared to 110. That's why a whisper can be heard easily in a quiet room, but you have to shout to be heard at a noisy party. (We deal with many issues in terms of rate, rather than absolute value every day – think about salary raises, interest rate, mortality, etc.)

Keeping the Weber–Fechner law in mind, we can unravel the peacock tail's extravagance – a story of evolutionary showmanship escalating out of control. Way back then, all peacocks had modest little tails, and females had no trouble spotting the flashy males who were just slightly better endowed. But as shorter-tailed males fell out of fashion – and existence – the average tail length crept upward. Soon males needed increasingly showy feathers to rise above their rivals and catch a female's eye, fuelling a spiralling arms race in tail length. Think of it like compound interest: tails grew longer, faster and flashier with every generation. Eventually, the cost of lugging around these oversized tails – burning extra energy, struggling to fly and practically waving a sign at hungry predators – became so high that peacocks simply couldn't afford to grow them any bigger. Evolution finally threw up its hands and said, 'Enough is enough!'*

This reasoning prompted researchers in Mike's lab to test whether pre-existing female preferences could shape the evolution of male mating calls in túngara frogs. Their results

---

* Animals that carry large handicaps can be especially vulnerable to extinction. The Irish elk, for example, may have gone extinct partly because it couldn't adapt quickly enough when the environment changed.

show that, when presented with two calls from different males, females chose the ones with more chucks based on the relative – not absolute – scale. Remarkably, the frog's predator, the frog-eating bat, also reacted similarly, sizing up prey on the relative difference in chuck numbers. Both results aligned perfectly with the Weber–Fechner law.[41]

So what's the big deal about this study? It tells us something pretty surprising: runaway evolution – like the wild spiral that gave peacocks those outrageous tails – doesn't necessarily require that the male trait and the female preference be genetically linked, as Fisher assumed. Nor does it have to show any male qualities or 'good genes' involved as Zahavi theorized. Instead, a pre-existing female taste alone might be enough to push males towards flashier, riskier features over time.

Behind all these discoveries is Mike's unrelenting passion for science – much like Sally Otto, Curt Lively, Marion Petrie and other trailblazers unravelling the mysteries of sex. Science isn't just Mike's career – it's his entire life. His home is a living, breathing hub of intellectual activity: part backup lab for research, part salon for brainstorming new ideas, part scientific networking hotspot, and part party venue where triumphs are celebrated – sometimes with his favourite local rock bands jamming along. This dynamic and fun environment has become an incubator for world-class evolutionary biologists like Gil Rosenthal, Steve Phelps and Ingo Schlupp.

Austin was (and still is) a hotbed for mate choice research. Beyond Mike's Zoology (now Integrative Biology) Department, the Psychology Department was also buzzing with researchers tackling questions about mate choice.

People there – including Walt Wilczynski, David Buss, Judy Langlois and their students – would quickly become part of Mike's world of science, especially when it comes to humans.

. . .

When it comes to choosing mates in our own species, you've probably heard the classic line 'opposites attract' – the idea that we seek partners who fill in what we're missing. Or perhaps you've stumbled across Freud's claim that women chase after men resembling their dads, and men prefer women who remind them of Mom. Sure, these theories make for fun dinner-party conversations, but scientifically speaking they're still in need of solid data to back them up.

Human mate choice is incredibly varied, influenced by a mix of internal desires and external factors. Despite the complexity, though, we can still spot some general patterns. One of the most intriguing discoveries is the difference between how we choose partners for short-term flings versus long-term relationships. For that, evolutionary psychologist David Buss, a colleague of Mike's at UT Austin, has made a slew of interesting discoveries.*

He has found that, while both men and women make choices in relationships, women tend to be the choosier ones.[42] Short-term relationships can benefit women in several ways. They might offer immediate material rewards, access to men with supposedly 'good genes', the potential to develop into long-term relationships, or even serve as a backup when things go south with a current partner.

---

* Many ideas in evolutionary psychology face criticism, from accusations of flimsy science to overlooking diversity while making broad claims. Some of the findings discussed here remain contentious, and only further research will bring clarity to the debates.

Buss and many of his fellow psychologists have also discovered that women in short-term relationships, compared with those in long-term relationships, are more focused on the physical appeal of men. They pay more attention to traits that suggest virility, dominance and high social status – strong jaws, deep voices and athletic builds but, interestingly, not beards.[43] These features, they argue, often signal good genetic quality because they are usually linked to higher testosterone levels. The logic? Testosterone, as we know, can be rough on the body. Men who can handle high levels of it must have strong immune systems to fight off diseases and parasites. So in a way, masculine features are handicaps – the peacock's tail of the human species.[44]

In the context of long-term relationships and marriages, Buss and his colleague David Schmitt found that women often favour men who, rather than physical traits, exhibit social status, ambition, diligence and a commitment to family life – qualities like protection, love and kindness being especially valued.[45] Most notably, women across cultures tend to prioritize material security more than men do, seeking partners who come with stable finances, property and good earning potential. This isn't just a Jane Austen-era quirk; it's a reliable global pattern, popping up everywhere from hunter-gatherer tribes like Tanzania's Hadza to modern, bustling societies in the US, Europe and China.* Whether they're gathering berries or balancing a corporate spreadsheet, it seems women consistently prefer a mate with solid resources – no matter the century or continent.[46] Conversely, scholars

---

* This phenomenon is known as hypergamy – the tendency to seek a mate of higher social status or wealth than oneself. It is most often discussed in the context of female mating preferences. Notably, even when women achieve economic independence, this tendency persists.

like Laura Betzig have noted that financial instability is a common reason cited by women for seeking divorce.[47]

Compared to women, men are statistically more interested in short-term relationships with multiple partners – whether it's casual sex, one-night stands or friends with benefits – roughly aligning with Bateman's rule. They tend to be less choosy. Men are also more likely to engage in extramarital affairs and are more willing to sleep with total strangers. This pattern seems to hold true across different cultures as well.[48]

Perhaps the most provocative finding by evolutionary psychologists like Buss and Schmitt about male mate choice is men's preference for traits signalling youthfulness and fertility – facial features, skin tone, breast size, waist-to-hip ratio and overall femininity. Cross-cultural studies reinforce this trend, showing that men, on average, marry women about three years younger and often remarry even younger women after divorce.[49] However, these studies tend to be skewed towards men of means, leaving open the question of how wealth influences male mate choice – a level of complexity unseen in other animals.

Intelligence ranks highly in mate choice for both men and women, across short- and long-term relationships. This preference likely has played a significant role in the evolution of human cognitive capacity,[50] to survive a diverse range of physical and social environments. Psychologist Geoffrey Miller even suggests that the evolution of the human mind – and its creations like literature, art and humour – has been largely driven by mate choice.[51]

We should be cautious about making sweeping claims based on observations from today's Western societies, however, where individuals have the freedom to choose their partners. Historically, and in many cultures today, mate choice has not been a personal decision. As biologist Gil Rosenthal emphasizes in his 2017 book *Mate Choice*, marriage is often a social choice – a decision made by parents, sometimes involving other relatives and even the entire local community. In many situations, marriage is more of a social and economic transaction between families than a matter of personal preference, which takes us to the clash between individual desires and societal expectations in marriage. It's a theme that runs through a vast trove of literary works, reflecting the lives and societies of their times. From Shakespeare's *Romeo and Juliet* to the Chinese legend of *The Butterfly Lovers*, countless stories explore this conflict.

Despite the complexities in human culture, evolution has left its unmistakable mark on the part of our cognitive system used in mate choice. That's why we're drawn to masculine traits in men – like broad shoulders, strong jaws and rugged features – and feminine traits in women, such as soft facial contours and curves in all the 'right' places.[52]

A key consequence of selection, including sexual selection, is the reduction of variation. In mate choice this often appears as a preference for average traits, acting as a stabilizing force that smooths out deviations and homogenizes our preferences.* As a result, beauty becomes universal because our sense of it converges. This helps explain why someone deemed attractive in one culture is often found appealing in another, even vastly different, culture.[53]

---

* Darwin was acutely aware of the role of sexual selection in shaping human facial features. In *The Descent of Man* he observes: 'The faces of men and women, though they differ in general appearance, yet resemble each other in a multitude of minor details ... it is probable that sexual selection has played a part in modifying the features of the two sexes.'

A Taste for the Beautiful

Mike once invited Judy Langlois, a leading psychologist on human beauty, to speak in his graduate course on animal behaviour. She captivated the class by demonstrating how to create strikingly attractive faces simply by averaging facial images from ordinary people, like women students working in her lab. Most remarkably, she showed that even infants, just a few months old, prefer pretty faces. If beauty is universal, then so too is ugliness – a reminder that our minds are anything but blank slates – they come preloaded with biases.*

This prompts us to delve into the intricate workings of our cognitive system. Indeed, brain scans reveal that, compared to less attractive faces, the beautiful faces of the opposite sex activate several brain regions with names such as the nucleus accumbens, medial prefrontal cortex, dorsal anterior cingulate and orbitofrontal cortex.[54] These regions are deeply involved in processing pleasure, reward and decision-making: a neural response that helps explain why beauty holds such a powerful sway over us. Psychological studies show that men are willing to go to surprising lengths when an attractive woman enters the picture. They'll work harder even at boring tasks and take bigger risks, like pulling off flashier, more dangerous stunts on skateboards, all to impress a good-looking observer.[55]

Unfortunately, our instinctive reactions to beauty also have a darker side – they can make us biased. These built-in preferences, often originating from mate choice, fuel what's now popularly called 'pretty privilege', and this privilege

---

* Beauty-and-beast is a classical literary theme that appeals to us by transforming societal superficiality. That's why Victor Hugo's *The Hunchback of Notre-Dame* has been so successful.

doesn't just stay in the dating world; it leaks into everyday life, reinforcing stereotypes, feeding prejudices and ultimately paving the way for discrimination.

Though the saying 'beauty is in the eye of the beholder' holds some truth, society's concept of beauty is far from diverse. Our preferences are often channelled into a narrow mould – just look at any pageant show. To make matters worse, mass media saturates us with images of what's deemed 'beautiful', reinforcing these narrow ideals. It's as if pageant shows have seeped into our daily lives, shaping everything from job interviews to political campaigns.

If pageant shows teach us anything, it's that cognitive biases are a trap. They lure us into judging people by their looks – a psychological pitfall known as the halo effect – even in contexts where appearance is irrelevant. As a result, attractive women often receive more opportunities, better treatment and greater popularity with men. Meanwhile, age becomes a relentless adversary for many women, driving a growing dependence on cosmetic enhancements. From Botox to smooth wrinkles to liposuction to melt away fat, these quick fixes reflect the mounting pressure to conform to idealized societal standards of beauty.

For men, one of the most influential biases revolves around height. Tall men are often favoured as leaders, whether in the corporate world or politics, irrespective of their actual abilities. They tend to command more respect, earn higher salaries and receive more opportunities. In fact, every extra inch in height translates to a 1–3 per cent increase in earnings – a phenomenon known as the 'height premium' in economics. This bias comes at the expense of shorter men, who are frequently overlooked, denied opportunities and subjected to unnecessary stress and pressure.

A Taste for the Beautiful

The American presidential election illustrates how the halo effect of height and perceived virility can bias our choices. In the early years, before photographs became commonplace in newspapers, shorter candidates had a fair shot – presidents like John Adams (5 foot 7 inches or 170 centimetres) and James Madison (5 foot 4 inches or 162 centimetres) were elected partly because their stature wasn't on full display. But once visuals entered the mix, the odds for shorter candidates began to dwindle.

The real game-changer? Televised debates, where height becomes impossible to ignore, turning elections into something like political beauty pageants. Since the 1960s, taller candidates have consistently had the upper hand. That might explain why Marco Rubio showed up to a 2017 Republican primary debate in high-heeled boots – a bold move that quickly became a punchline. The irony is rich: America's greatest president, Abraham Lincoln, was a towering giant in stature but hardly a looker,* while one of the least competent, Warren G. Harding, had both height and matinee-idol looks.

In this chapter, we've seen that beauty is real – and in the animal kingdom, it's often the result of evolution by mate choice. But in our own species, beauty is a double-edged sword. On one hand, it opens doors. On the other, it slams them shut. The trouble with beauty is that it's skin-deep and always racing against the clock. Even the most dazzling faces and bodies can't outrun time. What's stunning at twenty might be invisible at fifty. And very few of us are lucky enough to have it all – brains, charm and looks. Most of us

---

* Lincoln was keenly aware of his homely appearance. When Stephen Douglas accused him of being 'two-faced' during a debate, Lincoln shot back 'If I really had two faces, do you think I'd be wearing this one?'

Figure 7.7. Abraham Lincoln (left) and Warren G. Harding (right)

fall short in one way or another, and life throws curveballs: an accident, an illness, a scar, a change in weight.

Even if everything goes right, age eventually strips beauty down to the bone. Who wants to be judged for being too short, too tall, too fat, too thin, too wrinkled, too young, too old, too sick, too disabled? Yet let's be honest – how many of us *don't* size people up at first glance, led by a quiet, gut-level pull towards what we think is beautiful? That's the paradox. Beauty can lift us up – or quietly, cruelly, leave us behind.

Zoom out for a moment. What do we see? Thanks to sex, evolution didn't just shuffle genes – it cranked up the drama. Sexual selection burst on to the scene and built a world full of love songs, flashy dances and ridiculous body parts. And yes, it also stirred up trouble. We've already got a sneak peek at the tension between males and females in the wild dating game. But that was just the warm-up. The real showdown – the full-blown battle of the sexes – is just getting started. Buckle up. The next chapter brings the fireworks.

8

THE MATING GAME

Sexual Conflict

When Gustave Flaubert published his debut novel *Madame Bovary* in 1856, he didn't expect to be dragged into court by the French government on charges of obscenity. But the trial, to his surprise, became the work's best advertisement. After he was acquitted, the novel became a bestseller. So just how scandalous was the book? Raunchy enough to warrant its author being threatened with jail time? You be the judge – here's a quick rundown of its plot.

Emma Bovary dreams of passion, luxury and a life far beyond her dull country marriage to the kind but clueless Charles, a local physician. Fuelled by reading romance novels and her own restless fantasies, she leaps into a series of doomed love affairs and wild shopping sprees, trying to fill the aching void inside her with silk dresses and sweet nothings. But reality refuses to play along. Her lovers are cowards, her debts spiral out of control, and her small-town world starts closing in. As the glitter fades and bills pile up, Emma finds herself trapped, not by villains but by her own desperate longing for more – more love, more beauty, more *life*. In the end, she swallows poison in a tragic bid for escape, leaving behind a husband who never saw it coming and a daughter who'll pay the price.

On the Origin of Sex

These days, when novels overflow with suspense, plot twists and even murder, a tale of extramarital affairs barely causes a ripple, let alone a scandal. And yet marriage – whether unfolding in a sleek modern city or a sleepy French village in the 1850s – remains both a sanctuary and a battlefield. At its heart lies an age-old tension: the clash of wants, needs and expectations between men and women. It's a conflict that time hasn't tamed, and one that still resists any neat or final resolution.*

Great fiction writers are often sharp-eyed students of human nature, and Flaubert is no exception, to say the least. Like many literary giants before and after him, he zeroes in on a timeless paradox: the tension between our private desires and the socially sanctioned mould of monogamous marriage. More than a century and a half later, *Madame Bovary* still serves as a powerful lens for examining the deeper undercurrents in sexual relationships. But where do these tensions actually come from? And just how far back – across species and evolutionary time – do they go?

We've already tackled the first question in earlier chapters, diving head-first into the age-old battle of the sexes – a clash sparked by evolution in making eggs big and sperm small. Now it's time to take on the second one: what happens *after* this split? How does the tug-of-war between males and females play out in the natural world?

• • •

For French readers in Flaubert's time, *Madame Bovary* may have felt uncomfortably real – a too-close-for-comfort portrayal

---

* For the purpose of illustration, this discussion will focus on normative heterosexual marriages rather than same-sex unions.

of the quiet clash between Emma and Charles Bovary. But for biologist David Barash and his daughter Nanelle, that age-old tug-of-war between partners is more than just literary drama. In their cheekily titled 2006 book *Madame Bovary's Ovaries*, they argue that male–female conflict isn't merely a familiar twist in love stories – it's moulded into our biology. From the tangled affairs of Greek gods to today's swoony bestsellers, the tension between what he wants and what she wants – out of life and each other – keeps the drama alive. But here's the kicker: it's not just us. Evolution wired this male–female mismatch in interests into everything from strutting peacocks to scheming primates, as we illustrated in the last chapter. The clash, though manifesting itself in cultural practices, is also nature's doing, written into the rules of life itself.

As we've seen in earlier chapters, the evolution of sex gave rise to two main players in the game of life – males and females – whose strategies often pull in opposite directions. To truly grasp how sexual conflict unfolds and where male and female interests collide,* we need to peek into the private lives of other animals, where this evolutionary tug-of-war plays out in vivid, often surprising, detail.

According to Bateman's rule, polygyny gives males a shot at sky-high reproductive success, but for most that's more fantasy than fact. The reason is simple maths: if every male tries to claim two or more females, there just aren't enough

---

* Biologists sort sexual conflict into two main forms. *Intralocus conflict* happens when the same genes are shared by both sexes but work better in one than the other. *Interlocus conflict* is more like an evolutionary tug-of-war, where traits in one sex evolve to benefit themselves but hurt the other, prompting a counter-response – like males coming up with ways to control mating and females developing resistance. (See Arnqvist, G. and Rowe, L., 2005. *Sexual Conflict* (Princeton, NJ: Princeton University Press.)

to go around. That makes mating a zero-sum game from a pure biological perspective. When a few males hit the jackpot, most are left empty-handed – much like the infamous mutiny community on Pitcairn Island, where unchecked male mate competition led to chaos and violence.

In mammals like red deer and elephant seals, males burn bright and fast – they spend years bulking up, but once they reach their reproductive prime it's a brief, intense blaze before they crash out of the mating arena. Females, generally, and by contrast, play the long game: they start reproducing earlier and keep at it for years. This stark difference in life history of these species shows how males and females can evolve dramatically different strategies to boost their Darwinian pay-off. And once again, what works for one sex often spells trouble for the other. That's why mating systems are rarely permanent fixtures. They're more like temporary truces – brief pauses in a never-ending tug-of-war between the sexes. If you want the perfect symbol of that uneasy peace, just look at monogamy.

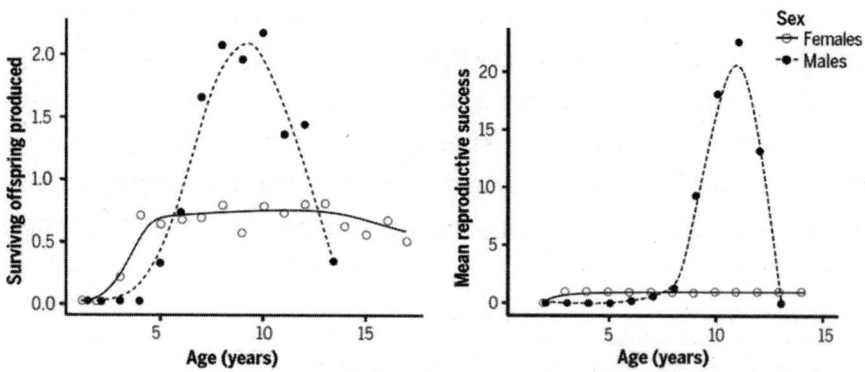

Figure 8.1. Sexual conflict often arises because males and females achieve reproductive success in different ways in many mammal species. Here are two examples: the red deer (left) and the elephant seal (right)

The Mating Game

Since animals aren't bound by marriage contracts like humans, monogamy seems like a real head-scratcher, or as biologists characterize it, a 'conundrum'.[1] Apparently, it's the worst way to settle sexual conflict. And yet, against all odds, it exists – and not just for humans, either. Why? Because certain ecological conditions make it worthwhile for both males and females to form one-on-one partnerships. For example, when mates are few and far between, it pays for a couple to stick together and pitch in with parental duties.* No roaming prospects, no wandering eyes – problem solved ... for the time being.

This is exactly what happens in the dwarf seahorse we met earlier. Genetic analysis reveals that males and females form a strictly monogamous pair – an oddity in the animal kingdom, though not unheard of. It turns out that, in the sparsely populated underwater world they inhabit, it's simply not worth the effort for both males and females to shop around for other mates.[2] A lonely sea leads to loyalty, you might say.

Now shift your focus to another fish – the princess pygmy goby. Living in shallow, brackish waters in the western Pacific Ocean, these fish often face a brutal reality: their eggs are a buffet for any passing predator. With such high stakes, parental duties become non-negotiable. Males, ever vigilant, don't dare stray from the nest to chase after new romantic

---

* In mammals, social monogamy appears to have evolved from a solitary lifestyle of females. (See Lukas, D. and Clutton-Brock, T.H., 2013. The evolution of social monogamy in mammals. Science, 341(6145), pp. 526–30; van Schaik, C.P. and Kappeler, P.M., 2003. The evolution of social monogamy in primates. In: U.H. Reichard and C. Boesch (eds), *Monogamy: Mating Strategies and Partnerships in Birds, Humans and Other Mammals* (Cambridge: Cambridge University Press), pp. 59–80.

prospects. Meanwhile, the females watch over their partners like vigilant bodyguards, ensuring no other female fish gets too close. The result? Monogamy – born out of necessity rather than loyalty.[3]

Even though some animals give monogamy a whirl, it's still relatively rare for a male and female to stick together as a couple. The animal kingdom just doesn't often offer the cosy ecological conditions needed – unless you're a bird.

For ages, birds, especially songbirds, were seen as the gold standard of fidelity – 'True as a wren at dawn', or 'Bound like bluebirds in spring', among other such idioms, abound. But that squeaky-clean reputation got a little muddy in the 1980s when genetic paternity tests started airing their dirty laundry. Today, we know that around 90 per cent of so-called 'faithful' bird pairs are dabbling in some extracurricular activities. Yep, sexual cheating is practically a pastime for birds.[4] In fact, studies found that out of 255 bird species, more than three in four (76 per cent) have chicks fathered by males other than their social partners. And nearly one in five (19 per cent) of offspring are fathered by extra-pair males.[5] Because of all this avian hanky-panky, biologists now draw a line between 'genetic monogamy' – for birds that actually stick to one mate – and 'social monogamy', where couples are more flexible in their commitments.

Scientists know that female animals aren't shy about playing the field. In fact, nearly 90 per cent of animals, not just birds, see females mating with multiple males.[6] You might be thinking, 'Okay, I get why males would want as many partners as possible, but what's in it for the females?'

Well, there are a couple of perks, all linked to the advantages of sexual reproduction. First, by mating with multiple males, females can increase their chances of finding a match with genes that complement theirs. Second, it's a way to hedge their bets. Instead of putting all their eggs in one genetic basket they spread the risk, which can lead to stronger, more diverse offspring with robust immune systems,[7] therefore maximizing the chance that their offspring live to adulthood and reproduce.

There's plenty of solid evidence that females often benefit from mating with multiple males, even if the exact hows and whys are still up for debate. In the grey foam-nest treefrog, for example, tadpoles stand a better chance of survival when the mother mates with ten to twelve males rather than just one.[8] In the dark-eyed junco, offspring from a female's extra-pair escapades tend to survive longer and are more successful in reproduction.[9] And let's not forget the superb starling – females here are all about extra-pair action, and the offspring from these affairs end up more genetically diverse than those conceived in the nest.[10]

Females don't treat all males equally, even when they're playing the field. They've got some tricks up their evolutionary sleeves. When a particularly attractive male struts into view, females often crank up their parenting game after mating with the better mate – laying bigger eggs or serving up gourmet meals for their chicks.[11] The collared flycatcher provides a prime example of this selective favouritism. Female flycatchers swoon for males sporting a bold white patch on their foreheads, the avian equivalent of a flashy sports car. But if her mate shows up without that signature style – that is, less attractive for her – she won't hesitate to sneak off for a little extracurricular romance.[12]

These are some cases where savvy females make it clear: even when juggling multiple mates, they're running the show, mostly without the knowledge of males. Every move is calculated to serve their own best interests, proving that behind the scenes they're quietly but firmly calling all the evolutionary shots.

Parental care isn't just a side gig in sexual conflict – it's another full-on battleground. Sure, we've already seen that male seahorses and jacanas are perfectly fine playing house dads, largely because there's not much else they can do to boost their fitness. But for most other creatures, convincing the male to stick around and help with the kids is like pulling teeth. (Remember Bateman's rule?)

Even though offspring benefit from both parents pitching in under most ecological conditions, males may only step up if they're sure the kids are actually theirs. Females, in response, may also tweak their reproductive strategies depending on how much effort the males put in. The burying beetle (*Nicrophorus vespilloides*) gives us a great example for this. In this species, if a male is confident those larvae are his, he'll share more of his precious carcass stash with them and eat less for himself. Sensing that her mate is all in, the female then lays smaller eggs, banking her resources for round two of baby-making.[13]

Certainty about paternity – knowing for sure those kids are yours – is a major incentive for dads to roll up their sleeves and get to work. That's why paternal care shines brightest in species where fertilization happens out in the open, like bony fish and amphibians. Here males can literally watch their sperm meet the eggs, leaving no room for doubt. And in this watery world of clear paternity, you'll find some truly heartwarming tales of fatherly dedication.

In common Surinam toads, for example, males carry their offspring on their backs like tiny, wiggly hitchhikers,

The Mating Game

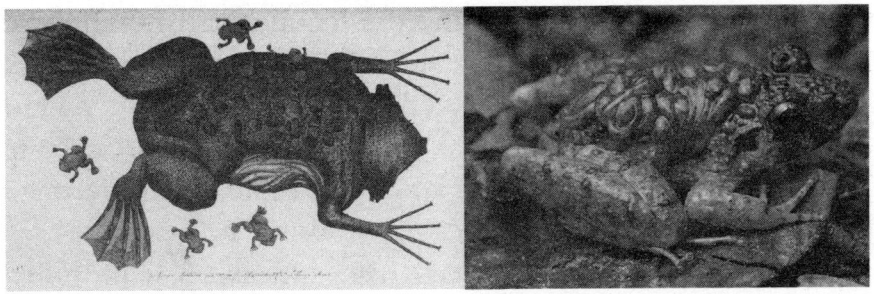

Figure 8.2. Males in Surinam toads (left) and smooth guardian frogs (right) carry tadpoles

turning fatherhood into a full-contact sport. Meanwhile, over in Borneo, the male smooth guardian frog doesn't just guard his eggs – he goes full chauffeur. Once the eggs hatch, he hoists the tadpoles on to his back and ferries them to safer pools, all while keeping predators at bay. These devoted dads are so committed to their little ones that even when fresh mating opportunities come knocking they resist the call.[14]

The simple fact that males can see their sperm connect with the eggs drives their paternal instincts in species with external fertilization. But in animals with internal fertilization – sharks, rays, reptiles, birds and mammals – things get trickier. Mating doesn't guarantee fertilization, and that gap gives females a chance to, well, browse for better fathers. Without clear visual confirmation of their paternity, this leaves males a little less enthusiastic about stepping into dad mode.

As we've mentioned, birds are the odd ones out when it comes to biparental care – both mom and dad usually pitch in. But male and female birds also engage in plenty of extra-pair flings, using them to settle sexual conflicts and sidestep the monogamy 'conundrum'. For this, the Eurasian penduline tit gives us a fantastic example. In this bird, the male goes all out, building a fancy nest,* and the female joins him for the big egg-laying moment. But then the trouble starts: who's going to sit on the eggs and feed the chicks once they hatch? Parenting means one bird has to stay behind. The catch? If one bails, he or she can go off and make more eggs or fertilize someone else's – great deal for them. But if *both* parents skip out, all that hard work on the current nest is wasted. And so begins the classic game of chicken. Who stays, who leaves, and how fast can they make a break for it?

It turns out that in about 30–40 per cent of nests, both parents ghost on the same day.[15] So who's more likely to desert first? It's usually the bird in better condition – the one who can afford to take the loss and start over somewhere else, leaving their partner stuck holding the nest, literally.[16] Once again, monogamy proves to be a tricky business because, in many cases, neither party is particularly thrilled with the arrangement.

This dynamic is even more striking in the spotless starling, a bird native to Spain. In this species, both parents typically share the task of feeding their chicks. But give the male a boost of testosterone and he starts acting like a hotshot – strutting, courting more actively, and slacking on parental care. Suddenly, he's too busy flaunting himself to bother feeding the young. Yet when his testosterone levels are reduced he returns to his parental duties, trading vanity for the daily grind of fatherhood.[17] This case, in addition to the pattern of attractive males often making lousy dads, highlights the powerful role testosterone plays in regulating the balance between mating effort and parental care in males.

---

* Male Eurasian penduline tits build a pouch-shaped nest that hangs from tree branches, often over water to avoid predators. For more security, the nest has a false entrance whereas a real one can be sealed off.

Figure 8.3. Eurasian penduline tits build complex and snug nests

Hormonal control of mating systems has been an enticing area of research in mammals, particularly in two species of voles: the monogamous prairie vole and the promiscuous montane vole. If you implant a gene coding for a neuroprotein (a protein serving as a hormone in the central nervous system) called vasopressin receptor A (which plays a key role in pair-bonding) from a prairie vole into the brain of a male montane vole, suddenly this previously wandering bachelor turns into a committed, monogamous father.[18] It's a wild transformation – and one that reinforces the importance of hormonal influence on mating systems. This gene implantation experiment shows that hormones aren't just chemical messengers – they can also be nature's mood ring, helping animals size up their surroundings and decide what mating strategy gives them the best shot at passing on their genes. Sometimes that means sticking with one partner. At other times it means playing the field. Either way, it's all about boosting their evolutionary score.

Hormonal influence aside, one environmental factor that favours monogamy is the threat of infanticide by unrelated males in many mammal species. This grim behaviour is especially common in primates, where males kill offspring to bring the female back into egg production – oestrus or ovulation, technically. To combat this, females in some species form long-term bonds with a male partner, essentially using monogamy as a strategy to protect their young.[19]

While monogamy can help reduce the risk of infanticide, it's far from a cure-all for sexual conflict. In some mammals – like lions and langur monkeys – male takeovers of social groups often trigger waves of infanticide, as incoming males eliminate the offspring of their predecessors to bring females back into oestrus. Female mammals respond with a range of counter-strategies: some form alliances with kin or current mates to defend their young.[20] In certain rodents, when infanticide seems inevitable, females may even terminate their pregnancies – a ruthless but effective strategy to avoid wasting resources on doomed offspring. And in over 130 species, females use multiple mating to confuse paternity. By casting doubt on who fathered the offspring, they turn would-be killers into cautious guardians, unwilling to risk harming their own potential young.[21]

Let's not forget that, besides infanticide prevention, multiple mating offers another key benefit: avoiding reproductive failure due to genetic incompatibility.[22] This strategy appears particularly effective in species like shrews and small rodents, where multiple paternity is common. In these species, a significant proportion of females mate with two or more males during each breeding season, leading to litters sired by multiple fathers,[23] apparently hedging their genetic bets to increase the likelihood of producing healthy, viable offspring.

Sexual conflict is also vividly expressed through sexual deceit, which can bypass mate choice and offer easier access to desirable mates. Indeed, as I detailed in my previous book, *The Liars of Nature and the Nature of Liars*, many animals engage in this sort of trickery. In numerous bird and mammal species, males often employ deceitful strategies to outwit and disarm the careful screening processes females use to choose mates. These tactics can range from faking signals of parental care and competitive ability to mimicking female characteristics. Meanwhile, females aren't entirely innocent either – they frequently engage in covert extra-pair copulations, securing genetic diversity or attractive sons while their social mate is none the wiser. Sexual chicanery, it seems, is a common tool in the evolutionary arms race driven by sexual conflict.[24]

When tricks fail, some males resort to coercion – the most brazen and aggressive form of settling sexual conflict. In chimpanzees, for example, males may sexually assault females, with any fecund female becoming a potential target for harassment.[25] This brutal strategy is also seen in orangutans, where young males may forcibly mate with females.[26] Biruté Galdikas documented such behaviour in her autobiography *Reflections of Eden*, where she described a disturbing episode of a young male orangutan assaulting a woman working at a field research station in Borneo. Even actress Julia Roberts had a close call during the filming of a documentary at the same site, nearly becoming a victim of a male orangutan's sexual aggression.

As you can expect, females have evolved a variety of tactics to fend off coercive mating attempts. In species ranging from the moorland hawker dragonfly to the European common frog, females have been known to 'play possum', pretending to be dead to escape sexual harassment from unwanted but pushy males. In social species, females often seek the protection of dominant males, using these high-ranking individuals as bodyguards against the unwanted advances of lower-status males.[27]

Perhaps the most dramatic and unsettling case of sexual conflict occurs in ducks, as Patricia Brennan and her colleagues have discovered. Although we all know fully well that ducks are non-moral, it's hard to feel comfortable witnessing several male mallards force themselves on a female – a common sight in the wild. Unlike most birds, male ducks have evolved a corkscrew-shaped penis – but females aren't defenceless. In a remarkable evolutionary countermeasure, female ducks have evolved vaginas that spiral in the opposite direction, complete with dead ends and twists that make it difficult for unwanted males to achieve successful fertilization. It's nature's version of a lock-and-key system – one designed to give females greater control over which males sire their offspring.[28]

This anatomical arms race likely unfolded over evolutionary time like a game of chess: females may have first evolved spiralled vaginas to counter male coercion, which in turn pressured males to evolve twisting penises that aligned with the vaginal structure. Females then further evolved by reversing the spiral or increasing its complexity, leading to even more extreme adaptations. This escalation didn't stop at shape; the arms race extended to size. In species like the Muscovy duck, for instance, the male's penis can grow up to 20 centimetres, while the female's reproductive tract is not only longer but also even more convoluted, illustrating how

The Mating Game

sexual conflict can shape some truly bizarre evolutionary outcomes as we see today.[29]

. . .

Sexual conflict doesn't just play out via sneaky behaviour and clever strategies. It digs much deeper – right down to the genes. Consider this: a newborn foal can weigh around 100 pounds, while the mother pony herself might only tip the scales at a few hundred pounds. Now, what happens if a tiny pony mom is mated with a much larger stallion? You'd probably expect her to have some serious birthing issues, right? But as it turns out, pony moms are armed with a genetic ace up their sleeves, a little trick called 'genomic imprinting' – inactivating dads' genes by adding a methyl group on the DNA. This clever genetic manoeuvre stops the baby from ballooning to a size that would spell disaster.* Here, the male–female showdown doesn't just stay in the ring of behaviour; it goes 'nucleic', battling it out in the DNA itself.

As we've already seen in this chapter, sexual conflict is an all-out brawl at every stage of making the next generation. It's there in mating rates, fertilization efficiency, relative parental effort, remating behaviour and female reproductive rate.[30] And thanks to sperm competition and cryptic female choice, the drama doesn't end when the sperm are delivered. Even after the eggs are fertilized, the battle rages on. Now it's not just the parents squaring off. It's a genetic tug-of-war between each parent's DNA inside the embryo.

---

* Sexual conflict through genomic imprinting in the embryo is present not only in placental mammals, but it is also in a very different group of organisms with something like a placenta: flowering plants.

The truth is that the two versions of a gene (alleles) at the same spot (locus) don't always play nicely – another example of Dawkins's selfish gene. What's good for one sex can be downright bad for the other. We can see this conflict of interest play out dramatically in wild sheep on the island of Soay off Scotland, where researchers have tracked the family lives and fates of every individual for generations. One allele at the *Rxfp2* locus gives males bigger horns, advantageous for winning mate competition. But another version of the same gene codes for smaller horns, which aren't as impressive in a fight but can increase the offspring's chance of surviving through harsh winters.[31] Similar findings have emerged in three-spined sticklebacks, thanks to a research team led by Florent Sylvestre. They identified two regions of the genome where certain genes give an advantage to one sex – while putting the other at a disadvantage.[32] In both cases there is a clear trade-off between the alleles.*

Let's return to the curious case of crossing a small pony with a large horse. Strangely enough, it's not horses but rodents that offer a surprising clue to how this odd couple manages to make it work. The poster child for this genetic tug-of-war between males and females is a gene called insulin-like growth factor 2, or *Igf2*. Both parents pass on a copy of this gene, but only the one from Dad is active in the embryo.[33] This isn't just a growth boost for the baby – it's a strategic push

---

* It is important to note that genes conferring an advantage under one set of conditions may prove maladaptive under another. Thus there are no universally 'good genes' that are beneficial across all contexts. For instance, in seed beetles, males with superior sperm competitiveness sire offspring that exhibit reduced longevity (Bilde, T., Foged, A., Schilling, N. and Arnqvist, G., 2009. Postmating sexual selection favors males that sire offspring with low fitness. *Science*, 324(5935), pp.1705–6).

from Dad's genes to draw more resources from Mom. The trade-off is that this extra growth can come at a cost to the mother's health, draining her ability to invest in future offspring. So she fights back, producing a counteracting gene, *Igf2r* (the receptor), which can silence that pushy *Igf2* with a chemical tag called a methyl group. This methylation is actually the addition of a methyl group ($-CH_3$) to DNA, a tiny chemical tweak that can shut a gene down. If you remove this silencing tag in female mice, the babies can balloon in size – proof that this parental arms race is written into the genome.[34] But why all this genetic warfare between the sexes?

We know that males, with their tiny gametes with little energy and nutrient storage, are basically reproductive parasites. They can only achieve their Darwinian fitness through females. Once their genes have hit the jackpot of fertilizing eggs, their next mission is to ensure those babies not just survive but thrive. The strategy is to get the offspring to grow as fast and as big as possible. As we've seen with red deer and elephant seals, males often live life on the edge, in

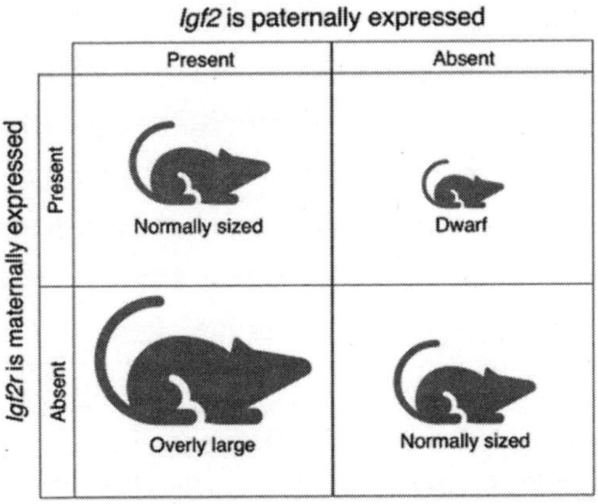

Figure 8.4. The size of an embryo is controlled by the paternal *Igf2* gene and maternal *Igf2r* gene

a Hobbesian world that's nasty, brutish and short. It's in their best interest to push females to go all-in on each offspring (at least, once they can be sure of their paternity). But for females, that's a dangerous game. The resources females have are like a fixed bank account: spend more now and you've got less for later. To play the long game, as most female animals have evolved to do, you can't go all-in. And let's not forget the mammalian twist – a baby that's too big can spell disaster during birth. As a result, the womb becomes a battleground for sexual conflict, where male and female spar with their genetic arsenals.

Turning off male genes is a female's secret defence strategy, protecting her from becoming the victim of male exploitation. This genetic shutdown can be a lifesaver, literally; it might explain why a Shetland pony mare can safely cross with a much larger horse stallion – such as a Shire – without ending up in serious trouble in giving birth.

It should come as no surprise that sex chromosomes tend to get the spotlight when it comes to sexual conflict. In mammals, for instance, some genes on the Y chromosome make males more attractive, but at a steep cost – offspring that are less likely to survive. (Similarly, in birds, genes on the W chromosome might boost the number of daughters, but at the expense of sons.[35]) Likewise, in guppies – a species with XY sex determination – some genes on the Y chromosome make males more colourful and attractive to females. But, as we know, because the Y chromosome does not recombine over most of its length, it tends to accumulate harmful mutations over time, falling victim to Muller's ratchet. As a result, these flashy Y-linked traits can be associated with reduced male survival. This presents females with a trade-off: mate with an attractive male and risk producing sons with lower viability, or choose a less ornamented male with potentially more robust offspring.[36]

Instead of wrestling with tough decisions in mate choice as guppy females do, some female mammals take a more drastic approach: they freeze up the entire sex chromosome – either X or Y – and shut down the operation of most genes there to avoid any discord. Still remember the chromosome test for sex/gender identification in sports we discussed back in Chapter 4? In females with two X chromosomes, only one is allowed to do its job while the other gets shut down, becoming a Barr body, the diminished existence of the sex chromosome. How does it work in animals?

Enter calico cats. In females, one of the two X chromosomes is randomly shut down in each cell during early development – a process called X-inactivation. This means that in some cells, the X from Mom stays active; in others, it's Dad's. Since the genes for fur colour – particularly orange and black pigments – are located on the X chromosome, this random inactivation creates a patchwork effect. The result is a feline masterpiece of orange and black spots, as if an abstract artist spilled paint on an exceptionally cute canvas.

But this randomness doesn't apply to all mammals. In marsupials and early-stage mouse embryos, the process is more one-sided: Dad's X chromosome is consistently silenced across the entire body – a phenomenon known as imprinted X-inactivation. That leaves Mom's X in charge.[37] Some scientists speculate that this maternal control could help explain why marsupial babies are born so tiny and underdeveloped – perhaps Mom's X takes a more measured, less growth-hungry approach, conserving resources for the long haul rather than splurging early.

Marsupials don't have a true placenta, but in placental mammals – also known as eutherians, including us – the placenta becomes ground zero for sexual conflict. It's a temporary organ, hanging around just long enough to see the pregnancy through – about thirty-nine weeks in humans. So who does the placenta actually belong to? The answer: the baby. It's an extension of the foetus's body, built from cells that carry two sets of chromosomes – one from Mom, one from Dad.

Now, here's where things get interesting in placental mammals. If you artificially create an embryo using only chromosomes from Mom *or* Dad, but not both, as biologists Davor Solter, Azim Surani, Sheila Barton and their team did with mice, you will end up with wildly different results. An embryo with only Mom's chromosomes produces an unusually small placenta, while one with just Dad's chromosomes ends up with a huge one.[38] So there is a clear tug-of-war going on: Mom's alleles 'want' to keep the placenta – and the embryo – on the smaller side, conserving resources for the future, while Dad's alleles push for a larger, more demanding placenta and embryo to maximize the chances of the embryo being born robust and strong and ready to hit the ground running. It's like they're arguing over the thermostat: Mom's trying to save energy, and Dad's cranking it up. In this genetic showdown, parent of origin matters a lot, with each side trying to tip the scales in its favour.*

We should remember that the placenta is just one of many arenas in the ongoing genetic battle between male and female mammals. Even after the babies are born, the fight continues. One such genetic battleground involves the gene

---

* You can read Abigail Tucker's *Mom Genes* (New York: Gallery Books, 2021) for more fun facts about human placentas.

Phlda2. In mice, only the mother's copy of this gene is active; Dad's version is silenced. When *Phlda2* expression is too low – say, if Mom's version is weakened – it can prime the female brain to go into overdrive, turning her into a supermom who obsessively nurses and licks her pups, often at the cost of her own health and future fertility. While it might seem like a win for the pups, this overzealous mothering actually benefits Dad – it's his sneaky genetic trick to wring extra care for his offspring, possibly at the mother's long-term expense.[39]

• • •

As we now know, male and female mammals tend to follow two very different playbooks when it comes to reproduction. Compared with modern societies, prehistoric men, much like red deer or elephant seals, tended to adopt a 'live-fast, love-hard, die-young' style with many meeting their end in brutal skirmishes. That's why men's bodies evolved for battle – strong muscles, high pain tolerance, and a physique built for a fight.[40] In terms of sheer strength, the average man can overpower all but one in 1,000 women.[41] Even today men generally seem less concerned than women about their health and how long they'll stick around.[42]

Female mammals, on the whole, tend to have stronger immune systems and live longer than males – a pattern especially pronounced in species with marked sexual dimorphism.[43] In humans, women generally invest in their reproductive fitness over the long haul. They reproduce during their prime years and, after menopause, often channel their energy into helping raise grandchildren, boosting their indirect fitness through kin selection. A longer lifespan translates to more opportunities to pass on genes, directly or indirectly through relatives. This evolutionary perspective may also help explain why women are more health-conscious and proactive about medical care – and why, despite often reporting more pain, they tend to tolerate it and survive longer than men.

When it comes to humans, mating systems aren't just driven by biology of course; culture, tradition and social institutions play a major role too. While many pre-industrial societies practised polygyny, monogamy also existed – and over time it became the dominant, legally sanctioned form of marriage in much of the world.[44] Polygyny allowed powerful men – emperors, kings, warriors, nobles – to accumulate multiple wives or concubines, often leaving men of lower status with few or no mating opportunities.[45] The rise of socially enforced monogamy was a turning point in human civilization. It wasn't just about love or romance; it was a tool for reproductive levelling, curbing elite male monopolies and promoting a more equal distribution of mating opportunities across the social ladder.[46]

Monogamy, as an institution, doesn't ensure reproductive equality. In practice, although many people are devoted to their marriages, still a good number engage in relationships outside formal pair bonds – whether discreetly or openly. For men with resources, extra-pair sex and remarriage can increase reproductive success.[47] Women, though less often, may also benefit when multiple men help raise their children.[48] These behaviours reflect underlying biological drives found across cultures. In Western societies, for example, about 20 per cent of women report infidelity.[49] Among the Himba of Namibia, extra-pair paternity reaches 48 per cent, compared to a global average of 4 per cent.[50]

Part of the challenge lies in the dynamics of mate choice. Many women are drawn to partners who combine physical appeal – often unconsciously read as signs of genetic fitness

– with the ability to provide stability and support. Yet men who possess both traits are relatively few, and often in high demand.* Those with qualities like attractiveness, confidence or charisma may attract partners easily, sometimes without offering long-term commitment.[51] This can create difficult choices. Some women navigate these trade-offs by adopting a dual strategy: seeking emotional fulfilment in one relationship while relying on another for long-term partnership and support.[52] Literary characters like Madame Bovary dramatize such tensions, which are echoed in real-world behaviour across cultures. Still, these strategies are shaped – and often constrained – by social context. In many societies, women who engage in multiple partnerships may face social stigma, financial disadvantages or, in extreme cases, threats to their safety.[53] Such risks highlight not just personal choices but the broader cultural frameworks that reward or punish them.

Female extra-pair relationships can lead to sperm competition, which helps explain why mate-guarding behaviours are found in many human societies – a point discussed in the previous chapter. Research shows that men may respond with increased possessiveness when their partners wear clothing seen as sexually attractive, such as red dresses compared to black ones.[54] Time apart can also heighten male sexual interest, possibly as an evolved response to perceived risk of infidelity.[55]

In some cases, concerns about fidelity can give rise to more troubling behaviours. Sexual coercion – an extreme

---

* Unfortunately, research shows that men with higher testosterone – especially those considered attractive – are statistically more likely to cheat and may be at greater risk of domestic abuse (Klimas, C., Ehlert, U., Lacker, T.J., Waldvogel, P. and Walther, A., 2019. Higher testosterone levels are associated with unfaithful behavior in men. *Biological Psychology*, 146, p.107730).

and harmful form of mate guarding – remains a reality in various cultural contexts.[56] While most societies formally prohibit such acts, enforcement and public attitudes vary, and in some places social norms may still downplay or overlook these violations. Understanding the roots of such behaviours does not justify them but can help illuminate the urgent need for cultural, legal and moral progress.

According to Bateman's rule, males in many species – including humans – can increase their reproductive success by mating with multiple partners. This can sometimes give rise to strategies that sidestep or undermine female choice. In both animals and humans, such sexual conflict may surface in forms ranging from subtle deception to coercion. In its most extreme and deeply unethical form it can lead to sexual violence.

While rape is universally condemned in modern legal and moral frameworks, history shows that it has occurred across cultures, particularly during times of war, displacement or societal collapse – when protections for women are weakest.[57] These disturbing patterns reflect an evolutionary tension between male strategies to maximize reproductive opportunities and female efforts to maintain control over their own reproduction. In humans, however, this biological conflict is overlaid with the powerful influences of culture, law and ethics – factors that hold the potential to restrain and reshape behaviour in ways no other species can.

Infanticide – one of the most disturbing outcomes of sexual conflict – has been documented in humans, especially when a male has no genetic connection to the child. This pattern has surfaced even in Western democracies. A landmark North American study in the 1980s found stark differences in abuse risk for children living with stepfathers compared to those with both biological parents. Among children from birth to age four, the risk of abuse was about twelve times higher; for those aged five to sixteen, it was five times higher. Although most step-parents are caring and devoted, abuse was far rarer in biologically intact families – occurring in only about 1 in 1,000 cases.[58] These sobering statistics reflect a biological vulnerability rooted in our evolutionary past. As Richard Dawkins once quipped, not without irony, 'We admit that we are like apes, but we seldom realize that we are apes.'

As we can all agree, there is a bottom line for humanity, and it is this: biology can help explain what we are, but it cannot dictate what we ought to do. Confusing the two commits the naturalistic fallacy – a serious error in moral and legal reasoning. This is especially crucial when discussing sexual conflict. The fact that it exists in nature does not – and must not – justify coercion or violence in any form. Evolutionary explanations describe behaviour; they do not excuse it.

Sex begets sexual conflict. In this chapter, we've journeyed through the wild terrain of battles between the sexes – from open physical struggles to subtle genetic skirmishes. But one big question still hangs in the air: where does all this conflict ultimately lead? To answer that we need to explore its evolutionary fallout. And for that we turn to the next – and final – chapter.

9

WAR AND PEACE IN THE BATTLE OF THE SEXES

Sex and Evolutionary Destiny

Let's play a game. Imagine you and your archrival, Alex, tearing down I-90 in a coast-to-coast showdown from Seattle to Boston, trunks loaded with top-secret packages so vital they could save the world, or at least land the winner a movie deal. This isn't an ordinary car race; it's chess on wheels too.

Those stakes aren't high enough? Let's fix that. Picture this: the freeway's got only two lanes: the *fast* lane and the *slow* lane. The fast lane? A white-knuckle thrill ride with no speed limits – pure adrenaline – but one wrong move and you're internet-famous for all the wrong reasons. The slow lane? Steady, safe and so slow it practically comes with knitting needles and an audiobook. You're cruising at a grandma-approved 50 miles per hour and forget about finishing first without a plan.

Now for the rules. You can drive only eight hours a day. Every hour, you must pull over, switch cars, reload the cargo and reset. No shortcuts, no loopholes, no excuses. When involved in a crash, wrecking your car, you are out. How do you outdrive Alex and claim victory in this 3,000-

mile showdown while ensuring that you and your cargo make it across the country?

The winning strategy seems obvious, right? Every time you switch cars, grab the fastest one with the lowest chance of turning you into roadside confetti. When you can't find a speed demon that's also crash-proof? Play it safe and cruise the slow lane. In short: chase speed when the risk is low and play defence when it's not. Nail the balance between pedal-to-the-metal and grandma mode, and you've got the best shot at leaving Alex in your rear-view mirror.

What's wild is this strategy isn't just a win for you – it's a win for your genes too. In fact, it's the name of the game for every creature that's ever rolled the dice on sex. Picture it this way: your body is your car – or the figurative Dawkins's vehicle used throughout this book. And those precious packages riding inside? That's your genes, along for the ride. But this isn't some Sunday cruise. Just like you, Alex, Sam and every other player in the race, your genes are gunning it – not down a freeway, but across generations. Their mission? To cross the ultimate finish line – our metaphorical Boston – before the competition does.

In this race, male bodies are the fast-lane cars: built for speed and risky manoeuvres. Female bodies are the slow-lane cruisers: safer and steadier. Rival genes hitch a ride on either one, hoping their host makes it far enough to pass them on to the next generation (the next leg of the trip). Sure, the real journey of genes is a tangled, twisty epic with more rules and no rest stops. But as you'll see in this chapter, the parallels between your freeway face-off and the evolutionary relay race are similar in many ways.

So how exactly do genes boost their odds of winning the evolutionary race by 'riding' with males or females? For a long time this question sat quietly in the corner – unanswered, underestimated and, frankly, underappreciated – until 1973.

Enter Robert Trivers, a sharp-thinking evolutionary biologist with a knack for flipping assumptions on their head. The breakthrough came not from a high-powered research lab but from a curious question posed by one of his Biology 101 students at Harvard – Dan Willard. Trivers took that spark and turned it into a short scientific paper that would go on to reshape how we understand sex, strategy and survival. The centre of the idea is misleadingly simple: sex ratio.

. . .

Back in the day – before the twentieth century – people thought sex ratios were like gravity: dependable, unshakeable and not something you questioned. Boys and girls showed up in roughly equal numbers, and that was that. But then, in 1930, Ronald A. Fisher, whom we met earlier, dropped a bombshell: sex ratios aren't carved in stone – they're a product of evolution's behind-the-scenes negotiations.

Here's the gist: if one sex becomes scarce, parents who produce more of that rare sex end up with more grandchildren. Too many guys? Daughters become the hot ticket. Too many gals? Sons suddenly seem like a smart investment. The rarer sex gets more attention, more mates and more chances to pass on their genes. So while it looks like nature always aims for a 50/50 split, what's really happening is a constant game of evolutionary tug-of-war. The 1:1 sex ratio isn't a hard rule – it's more like a temporary ceasefire, with natural selection quietly keeping the peace.[1]

To put Fisher's idea to the test, biologist Alexandra Basolo ran a clever experiment with platyfish (*Xiphophorus maculatus*). She purposely tipped the scales – creating fish tanks overflowing with either males or females – to see how the population would respond. The result? Nature didn't just sit back and watch. The sex ratios snapped back towards an even 1/1 split, like a rubber band finding its resting point. Sometimes they even overcorrected before settling down. Basolo's work showed that sex ratios aren't just balanced – they self-correct, thanks to the hidden gears of evolution quietly working in the background (Figure 9.1).[2]

So Fisher was spot on: when one sex is in the minority, it gets an edge in the mating game. That's evolution's way of evening the score. This built-in fix-it system – known as frequency-dependent selection – kicks in whenever the balance tips too far. It keeps the numbers from going lopsided for long. In the end, it's not about favouritism – it's about fairness. Males and females end up with equal evolutionary pay-offs, which is why sex ratios tend to hover around that magical 1:1.

There's a catch, of course: even in a population with a perfectly balanced sex ratio, not all males and females are created equal. Sure, on average, males and females end up

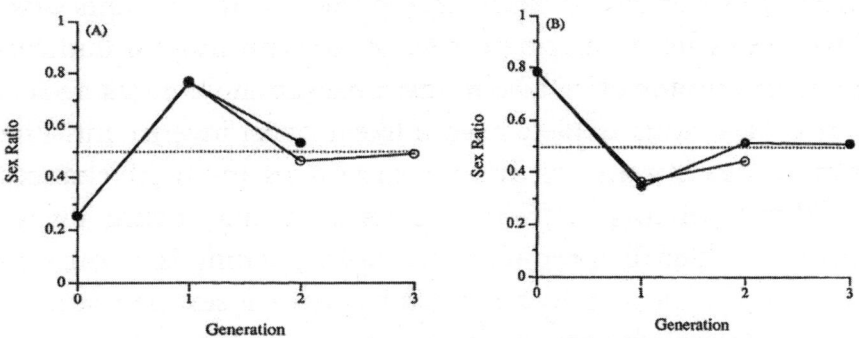

Figure 9.1. Sex ratio fluctuations in the platyfish over three generations after being experimentally distorted from the even split (0.50/0.50) to fewer males (left: starting with 0.23/0.77 as the male to female ratio) or more males (right: starting with 0.80/0.20 as the male to female ratio).

with the same evolutionary scorecard – Darwinian fitness – but that doesn't mean every individual should aim for a 50/50 mix of sons and daughters.

In species with polygynous mating systems, a few top-tier males hog most of the action while the weaker ones get left in the dust – often outperformed by females when it comes to passing on their genes. And since females usually invest far more time, energy and resources into their offspring, choosing whether to have a son or a daughter becomes a strategy. For a mom aiming to maximize her evolutionary success, she must read the room and stack the odds, tweaking the sex of her babies to get the biggest pay-off.

If females are indeed secretly rigging the sex ratio of their offspring, the best place to catch them in the act is probably among vertebrates – especially birds and mammals. Why? Because they're playing a high-stakes game where every kid counts.

Invertebrates like mosquitoes and mayflies take the scattershot approach: lay hundreds or thousands of eggs, toss them into the world and move on. No daycare, no dinners, no bedtime stories. Their populations rise and fall like rollercoasters, and in that chaotic lottery, tweaking the ratio of sons to daughters won't move the evolutionary needle much. But birds and mammals invest deeply in fewer young – feeding them, guarding them, even teaching them the ropes – in a slow-burn strategy where each baby is a major investment. This makes the decision of whether to produce a son or a daughter not just a biological coin toss but a strategic move in the game of Darwinian chess.

Another reason females might tinker with the sex ratio? They're gamblers – but smart ones who always keep a safety net. Take songbirds, for example. Most moms rely on dads to help raise the chicks, but that doesn't stop them from hedging their bets. Many sneak off for extra-pair flings, just in case their main guy turns out to be shooting blanks or singing lies (as we've seen in the last three chapters).

Female birds also use a similar strategy with their eggs. Some species hedge by laying eggs of different sizes – large ones that give a chick a solid head start, and smaller ones that are more of a 'why not?' bonus if the stars align. And then there's the kiwi, the ultimate high-stakes player. Kiwi lay a single, monstrously large egg – up to 25 per cent of their own body weight – then hand it off to dad, who incubates it for a punishing seventy-five to eighty-five days. If conditions are good, she may double down and lay another one. That's commitment roulette – with dad doing all the sitting.

However, when things take a turn for the worse, female birds rarely double down to rescue a failing investment. Instead they shift their focus to the healthiest chicks, tuning out the desperate cries of the weaker ones. Heartbreaking as it sounds,[3] it's their way of cutting the losses early and avoiding the trap of the sunk-cost fallacy. In the wild, survival is a cold numbers game, and moms instinctively know when to stop throwing good resources after bad prospects.

This cold calculation in how females decide to best manage their parental investment got Robert Trivers thinking: what if mothers could manipulate the sex of their offspring to boost their own fitness? While Fisher saw evolution as a reactive process, adjusting to gains and losses one generation at a time, Trivers took it a step further. He wasn't just thinking about the next generation – he was looking two generations ahead. This opens up a whole new perspective on the evolutionary landscape.

In polygynous societies, as we all know, male fitness is all over the map: a few star performers rake in the mating success while the rest are left in the dust. If a female can give birth to and raise a successful son, she hits the jackpot in the form of a wealth of grandchildren. It's a risky bet, but if it pays off it can be wildly profitable.

Female fitness, on the other hand, tends to be relatively steady. Even a daughter in poor condition – bad nourishment, poor health, low social status – is more likely to produce grandchildren than a weak son,[4] who might flop completely at reproduction – the metaphorical car crash, more often in the fast lane. So if a female in good condition can produce more sons than daughters, she stands a better chance of raising successful sons, whose victories could lead to an abundance of grandchildren. Conversely, a female in poor condition should play it safe, producing more daughters to ensure the survival of her lineage for now, while hoping for a future opportunity to boost its presence in the gene pool.

This idea, now known as the Trivers–Willard hypothesis (or effect), taps into the vast difference in reproductive potential between males and females, making it an essential part of reproductive strategy. It mirrors the tactic you might use in a car race with Alex – balancing fast moves with safe manoeuvres to win.

This armchair theorizing isn't just academic – it plays out in the real world across a variety of animals. In the brown anole, for example, females produce more sons after mating with large and attractive males, but produce more daughters after mating with small males, considered less attractive and less competitive.[5] The same can be said in the peafowl population Marion Petrie and her colleagues studied, where females in good physical condition lay eggs and hatch more sons.[6] In the common opossum, likewise, if you supply females with extra nutrition, such as dead fish, they produce more sons than daughters.[7]

It can't be stressed enough that the Trivers–Willard effect lies in a long game. It doesn't fully reveal itself in just one generation – you have to follow the story into the grandchildren's chapter. Especially in mammals, the benefits of skewing the sex ratio might seem invisible if you stop at the first set of offspring. But zoom out and the picture sharpens.[8] In fact, biologist Collette Thogerson and colleagues tracked ninety-year records of the reproductive success of 198 mammal species in captivity and found that tweaking the offspring sex ratio can lead to greater fitness than sticking with the fixed 1:1 ratio. And interestingly, male-biased offspring tend to outperform female-biased ratios in terms of reproductive success, although the reason isn't entirely clear.[9]

Sometimes, the Trivers–Willard effect seems to run in reverse. A great example comes from the Seychelles warbler studied by ecologist Jan Komdeur and colleagues. When times are tough and food is scarce, these savvy bird moms produce more sons than daughters. But when food is plentiful, they flip the ratio and favour daughters instead.[10]

The secret to this reversal lies in the warbler's family dynamics. Daughters stick close to home, inheriting the parents' territory and pitching in to raise their younger siblings – a big win for the parents' evolutionary score. Sons, meanwhile, are kicked out of the nest, left to fend for themselves and battle for scraps of territory elsewhere, often in rougher neighbourhoods. Sure, their future is a gamble – but sending them off also means fewer mouths to feed at home. It's a trade-off: high-risk sons with low expected returns but helpful in thinning out the dinner line, versus stay-at-home daughters who boost the family's long game.

Whether it's daughters in boom times or sons in busts, warbler moms are playing the same game: adjusting the sex ratio to squeeze out the most evolutionary bang for their buck.* Here, slower but safer on the slow lane may give you a better chance to reach Boston earlier.

So far we've been talking about sex ratios before and around birth, but those numbers can keep changing long after the babies are born. In many species, males are like fireworks – they burn bright but fizzle out fast. Thanks to fierce competition for mates, they tend to die sooner, leaving the adult population looking more and more female as time goes on.

But in species that flip the gender script, like jacanas, phalaropes and painted snipes, it's the males who stick around longer.[11] Here the females take on the macho role, battling it out for the chance to reproduce, which often leads to their early demise – just like the poor male red deer or elephant seals who kick the bucket after too many head-butting contests. Meanwhile the males, playing the long game, settle down to raise the kids and stretch out their lives, entirely upending the usual pattern where females outlast the males.

• • •

Now, when and how can the sex ratio be changed? For the answer, let's dive back into the world of vertebrates, especially birds and mammals, since most of the research centres on them. It turns out many species have tricks up their sleeves at every stage, from the moment they mate to when the babies are born, and even afterwards.[12] They're like

---

* What the Trivers and Willard hypothesis is saying is that mothers in good condition should invest more in the sex that can realize greater reproductive success. Therefore it's not always sons. In some conditions it can be daughters who can boost the mothers' fitness.

evolutionary stage managers, tweaking the line-up as circumstances change: after all, the show must go on.

There are many ways to meddle with the sex ratio – but at the heart of it, it often comes down to genes playing favourites. Most genes on regular chromosomes aim for a fair 50/50 split between sons and daughters, but genes linked to sex chromosomes are shamelessly biased. This is a textbook example of sexual conflict, as we explored in Chapter 8. In the lab mice, for example, nearly 600 genes are expressed differently between males and females in certain tissues. Why? Because it's a tug-of-war – genes fighting over whose agenda gets carried out and which sex gains the upper hand.[13]

The Y chromosome in mammals or Z chromosome in birds is all about Team Male and want to crank out as many boys as possible.[14] In contrast, the X and W chromosomes, running the female show, are just as biased, cheering for more girls.[15] Then there's the mitochondrial and chloroplast genes, passed down exclusively through females, who couldn't care less about males. For them males are dead ends, like cul-de-sacs with no way out. In fact, in some plants these selfish genes even cause male sterility, as though they work to ensure the females win.[16]

The X chromosome is a hotbed for the genetic battle of the sexes. With two copies cruising around in females and just one rattling solo in males, you have to wonder – does it play favourites? You bet it does. X-linked genes don't just sit quietly – they ramp up their expression in females, making sure they're heard loud and clear. It's like they're asserting their dominance, pushing female interests front and centre. When these genes end up in males they're less enthusiastic, almost as if they were saying, 'Sure, we'll help, but our real loyalty lies with the females.'[17]

How, then, do moms figure out whether it's better to produce sons or daughters? Turns out they may get the inside scoop from hormones – those trusty messengers that report on what's happening both inside and outside the body. In birds, the environment can influence these maternal hormones, which then get busy tweaking meiosis, the process that creates sex cells. Thanks to this system, avian moms might actually have the power to decide whether they send out a Z or a W chromosome, calling the shots on whether they'll hatch more boys or girls.[18]

In the rock pigeon, for instance, when a mother's testosterone levels spike, she's more likely to produce sons (ZZ). But if stress hormones are running high, she tends to have more daughters (ZW).[19] In some species of birds, females, tuned in to these same hormonal cues, don't treat all chicks equally. If the father is a handsome, high-stepping show-off, the mother will feed *his* chicks more generously than those sired by a less impressive mate.[20] As we brought up in Chapter 8, when a female mates with a flashy, genetically desirable male, she'll actually load her eggs with extra testosterone – giving her sons a hormonal jump-start towards being just as dazzling as their dad. But if she ends up with a dud, the eggs get no such boost. No VIP perks for average genes.

The same pattern shows up in mammals, where testosterone often acts as the hormone of high status. As we covered in Chapter 5, in species ranging from hyenas to caribou to shrews, a healthy dose of testosterone can give females the edge in physical battles, helping them claw their way to the top of the social ladder. But the story doesn't end there. Dominant females – those riding high on testosterone – are also more likely to tilt the reproductive odds towards sons. The theory goes like this: higher testosterone may influence the uterine environment or the timing of conception in ways that favour sperm carrying a Y chromosome, giving dominant moms a better shot at producing sons, who stand to inherit and benefit from that dominance.[21] This explains how, in species ranging from field voles to Nubian ibexes, high testosterone levels – especially before conception – are linked to a boy-heavy outcome.[22]

It's not just testosterone getting in on the action – other hormones tied to social status can also tip the scales in favour of sons. Consider progesterone. It's linked to female dominance, and surprise, surprise: moms with higher levels of it tend to pop out more boys.[23] But, as with birds, when there's a crowd of males around, female mammals – like lab mice – get flooded with stress hormones and suddenly the balance shifts towards daughters.[24] It's as if the moms are thinking, 'Hmm, too many boys in the room, better bring in some more girls!' In humans, stress hormones may have a similar effect, potentially causing male foetuses to be lost, leaving a higher ratio of daughters.[25]

We all know that nutrition is key to a female's health, but how does a mom's diet impact the sex ratio of her offspring? It seems there's a 'middleman' – blood glucose – involved, but the story's a bit more complicated than you'd think. High glucose levels, which signal that a mom is well fed and thriving, can be harmful for early-stage female embryos in mammals.* So when there's plenty of glucose in the system,

---

* A study on bovine embryo development reveals that male embryos are more likely than female embryos to survive when glucose levels exceed 2.5 mM. This difference in survival may stem from heightened activity in the pentose-phosphate (PP) pathway, which affects female embryos more negatively (Kimura, K., Spate, L.D., Green, M.P. et al., 2005. Sexual dimorphism in development of bovine embryos produced in vitro and its influence on the sex ratio of calves born after embryo transfer. *Molecular Reproduction and Development*, 72(4), pp. 494–501).

male embryos are more likely to thrive.[26] And when glucose teams up with testosterone, the result, you guessed it, is more sons. They're the ultimate chemical power couple when it comes to producing boys.[27]

It's not too surprising that females have ways to steer the sex ratio of their offspring – but they're not the only ones. Males are getting in on the action too, and studies suggest that guys can tweak their sperm strategy depending on how attractive their mate is or how intense the competition is from rival males.[28] In some species, males even adjust the proportion of X- versus Y-bearing sperm in their ejaculate, subtly stacking the odds. That's one reason dominant males across a wide range of species tend to sire more sons than daughters.[29] So yes, dads are pulling the strings as well, whether they're bugs, birds or mammals.[30] In fact, males are active players in the game of manipulating sex ratios to score an evolutionary advantage.* The battle of the sexes rages on in ways that push not only the limits of biology but also our imaginations.

. . .

How much do animals really stand to gain from manipulating the sex ratio of their offspring over time? Or, put another way, does cranking out more sons or daughters really make a difference in evolution in the long run? To answer that we must dive into a provocative concept called the 'sexy-son' hypothesis, cooked up by biologists Patrick Weatherhead and Raleigh Robertson.[31]

According to Fisher, when an attractive male trait – like the flashy orange spots on a guppy, a peacock's extravagant

---

* The biological mechanisms behind this process, particularly before offspring are born, remain largely unknown.

tail or the massive antlers of a bull moose – is genetically paired with female preference for that trait, it can trigger runaway evolution. The trait just keeps getting bigger, flashier and harder to ignore. But here's what Fisher didn't emphasize: when a female mates with a male who's a hit with many other females, she's not just getting a handsome partner – she's making a long-term investment in her genetic legacy. Her 'sexy sons' will have no trouble attracting mates, giving her a major fitness boost through her grandchildren. It's like playing the evolutionary stock market and hitting the jackpot.

The sexy-son hypothesis, in a nutshell, says that in a polygamous species, a hotshot son can unlock a massive amount of reproductive potential, leading to a big fitness windfall starting from the grandkid generation onward. This is a bonus that a 'sexy daughter' just can't quite match. So having a super-hot son is like holding the winning ticket in the evolutionary lottery. To put it in terms of your car race with Alex, it's like hitting the turbo boost and zipping past everyone in the fast lane on your way to Boston. Your sexy son isn't just keeping up – he's leaving the competition in the dust!*

If you feel that the sexy-son hypothesis sounds a bit like the Trivers–Willard hypothesis, you're not wrong – but they have different emphases. Trivers–Willard is all about the strategic balance of 'males for speed, females for safety' when it comes to parental investment. The sexy-son hypothesis, however, is laser-focused on one thing: tapping into the raw reproductive potential of male fertility. Forget about things

---

* This is called coalescence – an evolutionary term meaning that all versions of a shared trait (aka homologous trait, such as the mitochondrion or the Y chromosome) trace back to a single origin. In populations where more females than males consistently reproduce in each generation, that origin is more likely to come from a non-specific male than a female.

like paternal care, lavish nuptial gifts or prime territory.[32] In the world of the sexy son, a gene that manages to hitch a ride on a series of successful males can quickly outcompete rivals. For females who swoon over these handsome studs, it's like hitting the genetic jackpot. A sexy male isn't just eye candy – he's a powerful tool for boosting her genes in the gene pool.

Just how big of a boost can sexy sons give in the evolutionary race? Let's put it into perspective with a few jaw-dropping examples. In elephant seals, about 90 per cent of the offspring are sired by the dominant harem-holding males, leaving a measly 10 per cent produced by the majority of the unlucky males with no harems at all.[33] It's a similar story with Himalayan tahr – those blond, dashing males score 90 per cent of the copulations, while the darker, less flashy guys are left with the scraps – just 10 per cent.[34] And in the Tibetan macaques my colleagues and I have studied for over four decades, the alpha male runs the ultimate monopoly game during the mating season. He's not just the king of the hill – he's the emperor of amorous exploits, claiming up to a staggering 70 per cent of all copulations in a single season.* These cases show how reproductive success can skew wildly in favour of the 'sexy' males, giving their genes a massive leg up in the evolutionary race.

This 'home-run style' success for males isn't just about their reproductive effort – it's also powered by female promotion. You really see this in lekking species like grouse, manakins, topis and some bats, where females mostly flock to males with prime real estate in the centre of the territorial

---

* Recent genetic studies demonstrate that copulation rates do not necessarily correspond directly to fertilization rates because of extra-pair copulations; however, the two measures tend to be broadly correlated.

War and Peace in the Battle of the Sexes

Figure 9.2. A male Himalayan tahr (blond morph)

ground. In black grouse, however, the theme is played out with an additional variation: the males holding the prime spots in the centre of the lek change from year to year, but the same handful of studs still rake in most of the matings.[35] Why? Because older, more experienced females don't just choose the male in the right spot – they also judge his attractiveness. Younger, less experienced females, meanwhile, play it safe by copying the veterans. If the veterans are into a certain flashy male, the newbies follow suit, turning him into the latest feathered heartthrob. It's part merit, part popularity contest – and the result is a runaway effect where a few central males get all the glory while the rest just strut and hope. This mate-copying behaviour among females acts as the ultimate kingmaker, triggering a 'Matthew effect' in evolution: the rich get richer, with a handful of successful males raking in most of the genetic spoils, while the rest are left on the sidelines, merely spectators.

The sexy-son effect's fingerprints are all over the genome, thanks to strides made in DNA research. In polygynous bird species, the Z chromosomes – carried by males – evolve faster than in their monogamous counterparts, where things are a bit more even.[36] In humans, the story is even more striking. Genes passed down through men have evolved at roughly double the pace of those passed through women, as seen when comparing mitochondrial DNA (inherited from moms) with Y chromosome DNA (inherited from dads).* But this rapid evolution isn't without a hefty price tag. As history shows, it comes with a much steeper rate of reproductive failure and lineage extinction for men compared to women.[37] The sexy-son strategy might be a powerful engine for evolution, but it sure doesn't guarantee smooth sailing for everyone.

As you've seen, male and female fitness may balance out *on average*, but when you zoom out over hundreds or thousands of generations, the picture gets a lot fuzzier. That's because evolution isn't a tidy, predictable machine; it's a messy, stochastic process, full of random twists and turns. Even with a perfectly balanced sex ratio, not everyone gets an equal slice of the genetic pie. Some individuals get lucky

---

* Here, the 'rate of evolution' refers to how quickly genetic lineages accumulate changes that let us trace them back to a single common ancestor. And when we do that with the Y chromosome – passed only from father to son – we can track all modern men back to a common male ancestor who lived about 70,000 years ago. But if we follow the maternal line using mitochondrial DNA – passed from mothers to all their children – we land much further back, around 150,000 years ago. Why the difference? It's a reflection of genetic survival. Over time, more maternal lines have persisted than paternal ones. That's likely because fewer males reproduced successfully each generation, while more females did – leaving more continuous maternal lineages and making the mitochondrial tree stretch further into the past. In short: more moms left a legacy, while most dads were evolutionary dead ends.

and leave behind a genetic dynasty; most quietly fade out of the gene pool.

From this long-view perspective, strategies like the Trivers–Willard and sexy-son hypotheses start to make a lot more sense. They aren't just clever theories – they provide long-game insights into the subtle, often hidden tug-of-war between male and female strategies in the evolutionary arena.

Taking the long view of evolution gives us a new perspective on why so many males evolve such wild, over-the-top traits – purely to show off the attractiveness they will pass on to their sons. Never mind that the guys often contribute little else to their offspring. And yet females just can't seem to resist these flashy show-offs. Moreover, it also helps explain why females often 'play the field', opting for more partners than seems necessary to get the job of fertilization done.

This adds a fresh twist to how we think about male and female behaviour under sexual selection – even in examples we thought we already understood. Female pronghorns, for instance, don't just settle for the strongest nearby male. They'll go out of their way, crossing open terrain and dodging predators, just to track down the top stud.[38] In mallards, when a female mates with an especially dashing drake, she lays bigger eggs – as if doubling down on a high-stakes bet.[39] In some birds, as we've seen, females even sneak a little extra testosterone into their fertilized eggs – slipping in a secret booster to stack the odds in favour of producing sexy sons.[40]

But what about male mate choice, such as humans? There is an undeniable obsession with beauty – particularly female beauty – across cultures. The Ancient Greeks were so bewitched by Helen's face that they launched a decade-long war; some historians joke that if Cleopatra had a slightly crooked nose, Western history would've gone in a whole different direction; and we are all familiar with the so-called 'halo effect'.

This isn't just a Western phenomenon either. In ancient China, entire dynasties teetered on the edge thanks to the allure of beautiful women. Four legendary beauties – Xishi, Diaochan, Wang Zhaojun and Yang Yuhuan – were said to have changed the course of history just by walking into a room. Apparently, a pretty face could topple empires faster than a sword. But why *are* men so fixated on how women look?

Despite the magnetic allure of female beauty, science hasn't nailed down exactly why it matters. Sure, we've got some hints in the short-term stakes of a single generation – good health, good genes, higher fertility, lower parasite load – but none of these explanations seal the deal. Could beauty really be meaningless? It's highly unlikely. And, from a long-term perspective, an intriguing hypothesis emerges: if beauty draws men in, then beautiful women may have the upper hand when it comes to passing on their genes. Specifically, they might have a better shot at partnering with high-status, successful men – and producing those all-important sexy sons. In this evolutionary chess match, beauty could be a strategically safer move, boosting the odds of snagging a winner and staying in the game. While this idea is still speculative and lacks direct evidence, it fits neatly within existing theories of sexual selection and mate choice – and it's certainly worth exploring.

. . .

Does the Trivers–Willard effect apply to humans, you ask? Absolutely, and it plays out both biologically and culturally. If we consider places where polygyny is common, for example,

Rwanda, we can see that higher-ranking co-wives tend to have more sons than their lower-ranking counterparts, even when married to the same man.[41] Why? In these societies, sons often inherit a bigger chunk of the family wealth than daughters,[42] so having more boys could be a smart strategy for passing on status and resources.

Now compare that with the Mukogodo people in Kenya, who sit lower on the social ladder than groups like the Maasai and Samburu. Intermarriage happens, but many poor Mukogodo men stay single because they can't afford the bride price – money or goods paid to a bride's family by a groom for getting married. Even though polygyny is allowed, their chances of getting one wife – let alone more – are pretty slim. But Mukogodo women? They often marry into wealthier tribes. So the Mukogodos end up favouring daughters. Mothers tend to breastfeed their girls longer and take them to the clinic more often than they do with their boys. The result is that among Mukogodo children under five, the male to female sex ratio is heavily skewed at 0.67:1, with more girls than boys by over 30 per cent.[43]

Tampering with the sex ratio can happen subconsciously, especially in industrialized societies. Take Germany before the Berlin Wall fell: wealthier West Germany saw more boys born than the poorer East Germany.[44] In the United States, higher-income engineers have more sons, while nurses, who typically earn less, have more daughters.[45] In Britain, taller and bigger parents are more likely to have sons than their shorter, smaller counterparts.[46] Among the Forbes billionaires, 60 per cent of their children are sons, a much higher rate than in the general population. Interestingly, while women married to billionaires tend to have more sons, billionaire women themselves have fewer sons.[47]

You might feel excited or intrigued to see these interesting sex ratio data, but be aware that some studies don't find the Trivers–Willard effect at all. So the jury is still out. It suggests there are many factors at play, tangled up in ways we don't fully understand yet.*

The sexy-son effect shows up in humans too – though with a dark twist. Instead of heartthrobs and charmers, it often comes cloaked in violence. A striking 2003 study from a multinational team led by Tatiana Zerjal used genetic markers on the Y chromosome to uncover a staggering find. About 8 per cent of (16 million) men across a vast stretch of Asia – from the Pacific to the Caspian Sea – are direct descendants of Mongols from around 1,000 years ago.[48] Achieving such a genetic legacy would have been nearly impossible without the Mongols' relentless conquest of the world through sheer force and violence. Some media outlets even hailed Genghis Khan as the greatest lover in human history. The more realistic scenario is that many Mongol horsemen who shared a common Y chromosome contributed, forcefully, to the spread of the genes.

The Mongols weren't the only ones who spread their genes far and wide. A similar genetic legacy emerged in Latin America after European colonization began in 1492. Genetic studies consistently show that European Y chromosomes – passed from father to son – are widespread across the region, while Indigenous mitochondrial DNA – passed

---

* A recent study shows that there is a genetic foundation that some women tend to produce only sons or daughters. (Wang, S., Rosner, B.A., Huang, H., Rich-Edwards, J.W., Laden, F., Hart, J.E., Penney, K.L. and Chavarro, J.E., 2025. Is sex at birth a biological coin toss? Insights from a longitudinal and GWAS analysis. *Science Advances*, 11(29), p.eadu7402.)

War and Peace in the Battle of the Sexes

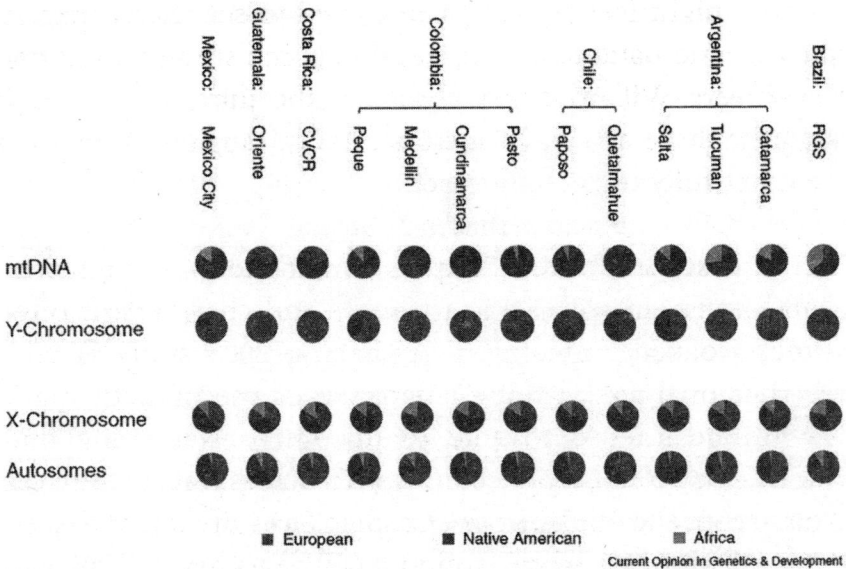

Figure 9.3. Genetic make-up in thirteen Latin American populations

from mother to child – remains prevalent. This asymmetry reflects a historical pattern in which European men fathered children with Indigenous women, often under conditions shaped by violence, coercion and profound power imbalances. Today, autosomal DNA (inherited from both parents) shows a more mixed ancestry, reflecting generations of admixture. Interestingly, genes on the X chromosome – of which women carry two copies and men only one – tend to show a stronger Indigenous contribution than the autosomes, highlighting the enduring genetic and cultural legacy of Native women across Latin America (Figure 9.3).[49] These genetic patterns line up neatly with the Trivers–Willard effect and the sexy-son hypothesis.

Genetics and history intertwine to reveal an uncomfortable truth: violence, as grim as it is, can sometimes pay off in evolutionary terms. Male aggression, though dialled down in today's societies, still leaves deep imprints. Anthropologists studying Indigenous groups lay bare the brutal legacy of village raids – men slaughtered, women seized as wives – a chilling echo of ancient patterns. As we explored in Chapter 6 through Steven Pinker's work, this isn't just ancient history. The cut-throat survival instincts of Stone-Age men still ripple through our veins. Even today, sexual jealousy remains a leading spark for homicides, whether in tribal communities or modern cities. Biologist David Barash doesn't mince words when he writes, 'If we could eliminate – or even significantly reduce – male violence, we would pretty much get rid of violence altogether.'[50] Hard to argue with that.

Violence tends to spike when the sex ratio tilts towards males. In polygynous societies like the Dogon of Mali, the Kipsigis of Kenya and the Aché of Paraguay, where men can have multiple wives, you end up with a lot of bachelors on the sidelines. This creates a sharp divide between the haves and the have-nots when it comes to reproductive opportunities.[51] The result? Intensifying competition among men for mates, and with it a greater chance for violence to erupt.

In many ancient societies, East and West, kings and emperors hoarded women in their harems while countless men at the bottom remained bachelors. And here's the thing – young single men are the most crime-prone,[52] making them a ticking time bomb for societal unrest. To keep the peace, these polygynous societies often relied on brutal punishments. This might explain why ancient laws were so harsh, as seen in Mesopotamia, the Vedic Valley, Mesoamerica and dynastic China. Even with these measures, chaos wasn't always avoidable – uprisings, revolutions and civil wars could still break out, especially fuelled by the frustration of those left out.

That's why reproductive fairness, like socially sanctioned monogamy, is so crucial – it keeps the sex ratio of the unmarried adults from tipping too far towards men. By levelling the playing field, monogamy becomes a key factor in maintaining societal stability and even fostering democracy. It's a biological reality that's often overlooked in the grand story of human civilization, but one that quietly shapes the very foundation of peaceful, stable societies.

Even with monogamy, however, an excess of men can still arise in situations such as male-skewed immigration or sex-selective abortion against female foetuses – a practice that's been common in parts of Asia, especially in rural India, China and South Korea. When the sex ratio leans too heavily towards men, the competition for mates rises steeply, leaving many men with slim-to-no chance of marriage. This imbalance can fuel a volatile social climate, often sparking a rise in male-driven crime – particularly violence.

Data in Britain and the US from the mid-nineteenth century to 1980 shows that violent crime rates shot up when men outnumbered women or marriage prospects shrank.[53] The same pattern holds in more recent studies in the US, where violence against female partners increases when there's a surplus of men.[54] In China, data from 1988 to 2004 showed that even a 0.01 increase in the sex ratio could trigger a 3 per cent rise in violent and property crimes. In fact, about one-seventh of the spike in crime rates can be traced back to this skewed sex ratio.[55]

But do all these dramatic battles over mates – especially among swaggering, showboating males – actually move the evolutionary needle? The answer: yes ... and no. That's because of one crucial genetic wildcard baked into the very definition of sex itself – recombination. This is where sex shows its real superpower: shuffling genes every generation like a deck of cards, making the future of any single lineage wildly unpredictable.

Recombination (specifically, crossing over between homologous chromosomes) is so crucial that it's worth a recap before you close this book. During the creation of sex cells – sperm and eggs – the chromosomes from the father and mother swap segments, mixing both male and female lines into a fresh combination (Figure 1.3 in Chapter 1). When this happens, the maternal and paternal lineages get all tangled up, and the evolutionary game resets.

To picture this in terms of your car race, we need to throw in one more twist: before switching cars each hour, you have to randomly swap some of your packages with Alex – or with anyone else who also joins the race. Do this enough times over the course of the journey, and by the time you roll into Boston you'll have no idea what's actually in your car. The cargo you loaded back in Seattle has been remixed so many times it barely resembles what you started with. That's recombination in action – making sure that even if you win the race, the genetic cargo crossing the finish line is barely predictable from that many generations ago.

Here's where it gets even more interesting: across the animal kingdom (and plants too) – from fruit flies to house sparrows to humans – geneticists have uncovered a 'nearly universal' pattern. Females, in most species, shuffle their genes more than males do.[56] It's like they've got a faster spin cycle in the genetic washing machine. Scientists have floated all kinds of hypotheses as to why this is – from the fine-tuned pressures of haploid selection and meiotic drive to the evolutionary push-and-pull of sexual conflict.[57] Yet despite decades of digging, the mystery remains.*

One clue may lead us to a potential new answer: when a chromosome, like a sex chromosome, gains an advantage, it dodges recombining with its counterpart – technically, homologous chromosomes – from the other parent, protecting its evolutionary lead. This raises a speculative possibility: recombination – when it happens, as in autosomes – can be tweaked to level the playing field, balancing how effectively males and females pass down their genes.

Imagine you're a woman in a made-up society of 200 people – 100 men and 100 women – all ready to have the next generation of 200 children, split evenly between boys and girls. On the surface, men and women seem equal in average fitness: together, they each contribute half the genes to the next generation. But sexual selection tilts the scale. Only fifty of the 100 men actually get to reproduce, while all 100 women do. So among those who *do* pass on their genes, each man, on average, has twice the reproductive success of each woman. That's a big imbalance.

Now imagine this keeps happening, generation after generation, and there's no gene-shuffling – no recombination. The long-term result would be that the DNA women carry would slowly be taken over by the genes of a few highly successful men and of eventually just one male

---

* Even the genomic landscape of recombination differs between the sexes: in males recombination events are predominantly localized near the tips (telomeric regions) of chromosomes, whereas in females they are more evenly distributed along the chromosomal arms.

ancestor.* Think Genghis Khan: one man's genes spreading like wildfire, pushing others out.

Except this isn't what actually happens, and the reason is recombination, nature's way of mixing things up. During reproduction, genes from both parents get shuffled, so the chromosomes you pass on are a patchwork of genes from both males and females. This keeps any one ancestor, usually a high-fitness male, from taking over the entire gene pool.

Think of it like our cross-country car race where everyone has to swap packages at random rest stops. By the time you reach Boston, only some of the packages in your trunk are the ones you loaded back in Seattle. The rest could come from Alex, Betty, Colin or whoever else joined the race along the way. Recombination works the same way – shuffling genes so that no one ancestor, no matter how successful, gets to dominate the gene pool for ever.

This constant genetic mixing helps level the playing field. It blunts the long-term advantage of a few high-fitness males and keeps genetic diversity alive. And because females benefit more from this process over generations – by preserving their own genetic contribution – it's no surprise that females have higher recombination rates than males in general. Here we see how different recombination rates can be used to resolve long-term sexual conflict between males and females

---

* Homologous traits – those inherited from a common ancestor – can be tracked down to a single origin through a genetic method called coalescence. This is the principle behind why all human mitochondrial DNA and Y chromosomes can be traced back to a single woman and a single man, respectively – often metaphorically referred to as mitochondrial Eve and Y-chromosomal Adam. Importantly, these individuals were not the only humans alive at their time, but rather the most recent common ancestors from whom these unbroken maternal and paternal lineages descend.

Although this idea from our thought experiment is still speculative, there's already some real-world evidence to support it. For instance, in Latin America, genetic studies show that while the Y chromosome – passed directly from father to son – is overwhelmingly of European origin, the autosomes (non-sex chromosomes inherited from both parents) still carry a strong Indigenous signal. This suggests that although European men dominated reproduction during colonization, Native American ancestry remains deeply embedded in the gene pool. One reason is recombination, which shuffles genetic material in each generation, especially in females, helping preserve Indigenous genetic heritage even as Y chromosomes were largely replaced.

Once recombination is understood as a sex-specific strategy shaped by sexual conflict, sex itself takes on a new evolutionary meaning. It's not just about dodging Muller's ratchet, repairing damaged DNA, combining good genes or staying ahead of parasites and pathogens. It's also a long-term genetic survival tactic – especially for females. By shuffling genes each generation, recombination helps preserve genetic diversity and keeps a female's lineage from being erased by a few overachieving males. From this strategic perspective, the 'safety' in the phrase 'females for safety' isn't just about avoiding risks – it also means protecting genetic legacy in the long run.

Now we see that while sexual reproduction can drive a wedge between the sexes, males and females are still tightly bound in their evolutionary fate through recombination. Despite all the fierce competition for mates and the high costs of mate choices, the outcomes of any current strategies, no matter how obvious they are in the short term, become less predictable as generations proceed. So after diving into the two epic battles – between the sexes and within each sex – we find both a truce and truth: in the long run, neither side truly wins.

As for the much-hyped 'battle of the sexes' in popular culture, it turns out it's more bark than bite when it comes to evolution. Looks like van Valen was right again here: all the drama and antics stirred up by sex are just the necessary evils of the Red Queen's game, only to keep species in the race. Males and females continue to exist as different vehicles for genes on their uncertain but entwined evolutionary journey. So next time, if someone claims one sex is superior to the other, just laugh – it's all part of the show, not science.

EPILOGUE

If there's one popular science book that defined our understanding of sex and gender in the pre-biotech era it's Matt Ridley's 1994 classic, *The Red Queen*. With razor-sharp wit and punchy prose, Ridley unravelled the cutting-edge science of his time, peeling back the layers of sex and evolution like a master curator. Reading it felt like racing through a scientific Louvre – every page a breathtaking exhibit on the mechanics of reproduction and selection.

But fast-forward three decades, and science has exploded like a supernova. Breakthroughs in molecular genetics, genomics, cell biology, developmental biology, neurobiology and biomedicine have expanded our understanding at warp speed. Imagine constructing ten new museums, each the size of the Louvre, overflowing with discoveries that defy intuition and stretch the limits of what we thought we knew about sex and gender. Could anyone write a single book that captures it all? Not a chance – science has outgrown any one volume.

That said, this book still aims to take you on a whirlwind tour. It's a fast, selective ride, but you've glimpsed some of the most electrifying discoveries – ones that push beyond the narrow slice of reality our senses can grasp. So before you close the book, let's take one last look at the journey we've travelled together — and a few gems to carry with you as intellectual souvenirs.

· · ·

Evolution is a game with one rule: you win by passing on more of your genes — fitness, as the ultimate currency of life — than your rivals. But don't be fooled by the rule's simplicity. The ways this game unfolds, shaped by an ever-shifting landscape of social and ecological pressures, make chess look like tic-tac-toe.

And then came sex — a game-changer of cosmic proportions. It didn't just add complexity; it unleashed chaos, spinning out wild, unpredictable strategies and dazzling emergent properties that no one, perhaps not even Darwin himself, could have seen coming.

To grasp the true significance of sex, we first need to step into a world without it — a world where life ran purely on cloning. At first glance, cloning seems like the perfect system: efficient, straightforward and free from the hassles of finding a mate. But beneath its simplicity lurks the fatal flaw of Muller's ratchet. Over generations, harmful mutations pile up like baggage with no return ticket, slowly dragging the lineage towards genetic ruin. Life needed a reset button. And then sex arrived — a genomic rescue mission, rewinding the ratchet before it could grind a species into oblivion.

But sex didn't come cheap. It brought with it a staggering price tag, known as the twofold cost: why invest in males — who can't bear offspring — when cloning could double reproductive output? This inefficiency poses a deep evolutionary puzzle, forcing us to ask what made sex worth it.

There's no single answer, but many scientists argue that sex first emerged as nature's master repair tool, particularly for fixing catastrophic DNA damage like double-strand breaks – with an extra perk: turning back Muller's ratchet. Others point to its power to shuffle genes, merging the best traits from different lineages – a genetic remixing that fuels adaptation and survival in an ever-changing world.

Just when we think we're closing in on the 'why' of sex, nature throws us another curveball: the presence of reproductive flip-floppers. These creatures – mostly asexual but occasionally dipping a toe into the sexual pool – suggest that the origin and the maintenance of sex are two separate riddles.

If sex is so indispensable, why then do some species treat it like an emergency backup rather than a full-time strategy? The deeper we dig, the clearer it becomes: the story of sex is far messier, more intricate and more surprising than we ever imagined.

As we chase down the reasons why sex persists, another revelation emerges: sex isn't just about repairing DNA or remixing genes – it's also a weapon in an evolutionary arms race. Here we meet the Red Queen hypothesis, which casts sex as a high-stakes survival strategy in the never-ending battle against parasites and pathogens. By constantly shuffling the genetic deck, sex gives offspring a fighting chance in this relentless tug-of-war. It's a game-changer, offering a powerful explanation of why so many species double down on sex as their primary mode of reproduction, but it's not the whole story – because some of the benefits are still not accounted for. So the search for the origin of sex continues.

By now, we can fully appreciate the profound role of sex in generating genetic diversity, thanks to the magic of recombination. This diversity unfolds on three levels. First, sex carves life into distinct reproductive roles – proto-sexes or fully fledged sexes – splitting organisms into playing male and female roles in most species. Second, it introduces two forms of existence – haploid and diploid – unlocking new ways to tune life cycles with remarkable flexibility. But perhaps most intriguingly, sex dismantles rigid genetic associations, reshuffling genes into fresh, dynamic communities.

Perhaps even more important, this genetic reshuffling is a kind of liberation. By breaking the bonds that once locked genes into fixed neighbourhoods, sex gives them freedom – a chance to move, mix and follow their own agendas. Each gene can now act in its own interest – sometimes teaming up, sometimes butting heads, and sometimes throwing the whole system into disarray. With sex, the genome becomes a bustling, unpredictable marketplace where genes strike deals, compete, manipulate, sabotage and hustle their way towards evolutionary success.

The genetic dynamism unleashed by sexual reproduction doesn't just stir up diversity – it also fuels tension within families. The moment kinship falls short of a perfect 100 per cent, competing interests creep in, setting the stage for conflict. This friction, known here as the social cost of sex, is an inevitable toll of genetic mixing.

But evolution, ever the problem-solver, introduced a counterforce: kin selection. Its mission? To ensure that genes shared among relatives get a helping hand, boosting their chances of survival and replication. This is why blood ties matter – why we feel an instinctive pull to protect and support our kin. It's not sentimentality; it's a time-tested strategy, written into the very fabric of life.

Kin selection unlocks a radical evolutionary strategy: organisms can remain in the genetic game even with reduced fertility – or without reproducing at all – by helping their relatives pass on shared genes. This cooperative approach stretches

Epilogue

the boundaries of reproductive success, making space for a spectrum of intersex as an adaptive feature of evolution.

From this perspective, the biological definition of sex – historically anchored in gamete size – demands a broader framework. Intersex individuals aren't anomalies; they are integral players in life's evolutionary chessboard, shaped by the same pressure – the pursuit of maximal fitness under a given set of conditions – that moulds all other reproductive strategies across species. The old dichotomy of sex is simply too rigid to capture the full spectrum of diversity and adaptability that evolution, ever inventive, has sculpted.

Sexual diversity doesn't stop at reproduction – it extends into the astonishing ways sex is determined and revealed during development. Along this journey in Part II, we met a cast of extraordinary creatures that settle the issue of sex identity through wildly different means: genetics (or chromosomes), environmental cues, and even social dynamics. But among these strategies, one stands out as the boldest move of all: sex reversal – a complete switch between males and females when conditions demand it.

Each of these mechanisms is a finely tuned adaptation, sculpted by evolution to navigate the specific ecological and social pressures an organism faces, all in pursuit of a singular goal – leaving behind more copies of the genes it carries. And in this intricate dance of sex determination and development, one truth becomes clear: variation rules the game. There is no universal blueprint, no rigid playbook. What we often call 'exceptions' are, in reality, the rule – a dazzling testament to evolution's boundless ingenuity.

While some species flip their sex identities, many more tweak their genders – shifting along especially the masculine-to-feminine spectrum – not out of whim, but as a calculated adaptive move. Here we met the undisputed gender-bending masters, the jacanas. The females swagger and compete like textbook alpha males, while the males take on the domestic workload – building nests, incubating eggs and raising chicks. Then there are spotted hyenas and European moles, who don't just swap gender roles but rewire their very anatomy, altering their reproductive organs to maximize reproductive success.

This reveals a deeper truth in gender's biological construction: gender is a flexible, adaptive strategy, woven from a rich tapestry of biological traits – morphology, hormones, neurobiology, genetics, behaviour – all fine-tuned for reproductive advantage under complex social and ecological conditions. And here's the kicker: while these traits are crucial, they are not absolute markers of biological sex. They shift, adapt and, like many things in biology, defy simple categorization.

Evolution, ever the rule-breaker, shatters moulds yet again when it comes to gender. It spins out variations so intricate, so unexpected, that they often defy the very language we use to describe them. The sheer diversity forces us to confront the edges of our understanding, making Ludwig Wittgenstein's admission all the more fitting: 'The limits of my language mean the limits of my world.' Nowhere is this truer than in the ever-expanding spectrum of gender – where evolution refuses to be confined by human categories, reminding us that reality is always richer – far richer – than the words we have for it.

Part III thrusts us back into the heart of the evolutionary battlefield, where males and females emerge as rival strategists, locked in an endless chess match under the relentless

Epilogue

force of sexual selection. Here the focus sharpens on how all bodies – male and female – are sculpted and optimized, not for survival alone but more essentially for reproductive success.

We stepped into the raw intensity of mate competition, where individuals of the same sex battle for the upper hand. This evolutionary arms race fuels the rise of jaw-dropping traits, driving sexual dimorphism in many species to extremes – from towering body sizes to extravagant weapons. But competition isn't just about brute force; it shapes intricate social dynamics, forging rules of engagement, alliances and even moments of cooperative bargaining – an evolutionary detente of sorts.

Beyond combat lies mate choice, an equally high-stakes game. Here, sexual selection has spun out a dizzying array of tactics, from flamboyant displays to psychological manipulations, all designed to tip the scales in favour of one suitor over another. Every peacock's tail, every intricate courtship dance, is a finely tuned advertisement in the marketplace of reproduction.

But sexual selection doesn't stop at crafting harmony between the sexes. It also ignites sexual conflict – a relentless tug-of-war where evolutionary interests diverge between males and females, sparking battles that range from physical confrontations to microscopic wars fought at the genetic level. Far from a mere struggle, this conflict fuels a cycle of adaptation and counter-adaptation, driving evolutionary arms races that shape the very fabric of life. Over time, neither sex emerges as the ultimate victor, yet both persist, endlessly refining their strategies in nature's longest-running contest – a perpetual, high-stakes game where the only true winner is evolution itself.

• • •

As we push forward, two enduring 'known unknowns' will continue to challenge us. The first is *demographic limitation*: as we uncover more cases and make more observations, the odds of stumbling upon new exceptions will only grow. Biology loves its rule-benders and rule-breakers, after all. The second is *technological limitation*: our understanding is always constrained by the tools of our time. As new technologies emerge, they'll undoubtedly take us deeper and further into the mysteries we've yet to solve. Science is always work in progress.

But despite these limitations, we're better equipped than ever to tackle many important questions with today's state-of-the-art knowledge. So let's wrap things up with a practical scenario – an exit quiz, if you will – to test how much you've gleaned from our scientific exploration. (Don't worry, no grades here – just a little fun to end on an unfinished journey.) Here is the scenario of the problem.

On 17 March 2021, another real event taking place during a US Senate Judiciary hearing on the Equality Act for LGBTQ Rights. The two actors are John Kennedy, Senator of Louisiana, and Alphonso David, President of the Human Rights Campaign.* Here is the transcript of their famous debate:

Kennedy: David, let me ask you a question. How many sexes do you think there are?
David: How many sexes? Well, there's a difference between sex and gender identity if that's what you're getting at.
Kennedy: No. I'm asking about biological sexes.

---

* Here is a YouTube video of the exchange: www.youtube.com/watch?v=nBoVdcXqKAQ.

Epilogue

David: Well, I would defer to the medical practitioners, but I think there have been studies showing that if you're talking about sex, sex is defined by many different characteristics and including chromosomes ...
Kennedy: Are there more than two?
David: You could make that argument that there might be.
Kennedy: Are you making that argument?
David: I can't ignore the fact that there're individuals who are intersex and so I'm just trying to ...
Kennedy: I'm running out of time. Are there more than two sexes in your opinion?
David: It's not limited to two.
Kennedy: Okay!

So who's right, scientifically speaking? (Spoiler: the answer might be 'both', 'neither', or 'a little bit of each'.) But if your answer is, 'Actually, I've got a better idea than either of them,' then you've just made my day.

Before you close this book, I want to share a final reflection on the journey of writing it. Sex, once it emerged, became a force unto itself – like a self-driving car that doesn't ask for directions but somehow keeps rewriting the map. Evolution tugs at the wheel, but the route remains largely unpredictable, only making sense in hindsight. Sexual reproduction opens so many doors that even after a lifetime of reading, researching and wrestling with the subject, I still find myself humbled by how much I don't know.

So if this book has surprised you – if you've come across things about sex that seemed almost impossible – I know exactly how you feel. Writing this has only deepened my awe at how little we truly understand. The mysteries of sex are no less profound than the deepest enigmas of the physical world – the Big Bang, black holes, dark matter. Sex, too, reshapes reality in ways we are just beginning to grasp.

ACKNOWLEDGEMENTS

If left to specialists, the story of the origin and evolution of sex might have stayed tucked away in the ivory tower of biology – debated among scholars but absent from everyday conversation. Yet discoveries in molecular, cell and developmental biology now arrive so quickly that even experts can struggle to keep up. Writing a book that is both rigorous and accessible turned out to be far more challenging than I imagined. Fortunately, an extraordinary group of colleagues, friends and family helped transform this project from a fragile idea into the book you now hold.

First and foremost, my deepest thanks go to my son, Orien – an exceptionally passionate student of the natural world with a scholar's grasp of molecular and cell biology. His expertise roams freely from viruses and bacteria to protozoa, fungi, plants and the small invertebrates that most of us notice only when they make their presence abundantly clear. He guided my choice of research materials, broadened my intellectual range and improved the manuscript from its earliest drafts to the final polish. His editorial touch shaped everything from the architecture of the book to the weight of a single word. I sometimes suspect that, had he been the author, the book might have been even better.

I am deeply grateful to those who generously shared their time in interviews: Mike Ryan, Curt Lively, Sally Otto, Marion Petrie, Joyce Benenson, Sarah Richardson, Matt Ridley and Richard Wrangham. Mike and Curt went further still, supplying key materials and sharp comments on early drafts. Gil Rosenthal offered invaluable feedback on Chapters 6 through 9, despite the demands of an unforgiving schedule and personal health challenges. Agustín Fuentes brought fresh perspectives to the chapters on sex and gender (4 and 5). Meaghan Boice-Green, Gabe Stryker and Lisa Norris helped shape the opening chapters with their essential guidance, while Patricia Greenfield and Sarah Hrdy enriched the work through thoughtful and inspiring conversations. And David Barash read the entire book, offering sound suggestions that helped me fine-tune the copy-edited manuscript.

The Hrdy Fellowship at the Harvard Radcliffe Institute for Advanced Study was a turning point – bestowing the rare gift of time and the stimulation of an extraordinary intellectual community. My fellowship year was brightened by conversations with brilliant and generous Radcliffe fellows, who expanded my thinking both inside and outside the seminar room (and taught me the subtle art of surviving – indeed enjoying – a karaoke party). I also benefited from lively exchanges with Harvard faculty and graduate students across disciplines, especially those in Martin Surbeck's Pan Lab. I remain grateful to Radcliffe's exceptional fellowship administrators – Claudia, Jimena, Sharon, Jin, Kayla, Alison, Jane and Maria – whose efficiency, kindness and perfectly timed bursts of humour made Cambridge feel less like an institution and more like a second home.

I owe a special debt to my remarkable Radcliffe research participants, Shayna Leng and Lucille Komar. With youthful

Acknowledgements

energy and unflagging dedication, they examined every chapter, paragraph, sentence and word, insisting on clarity, precision and coherence. Their rigorous critiques lifted the work immeasurably, while their friendship brought warmth and good humour to a long year. Shayna and Lucille, your contributions – and your company – remain among the brightest treasures of this journey.

This long journey began with my agent, Jessica Woollard, who saw promise in my idea and found it a welcoming home. That resonated instantly with my editor, Izzy Everington at Profile Books, whose cheerfulness, editorial acumen and steady guidance shepherded the manuscript from first sketch to final flourish. I am equally grateful to my meticulous copy-editor, Linden Lawson, whose sharp eyes dispatched more than a few embarrassing missteps, to Sarah Kennedy, who steered the production process with skill and grace, to Zara Sehr Ashraf for her help with image permissions and to Clare Sayer for proofreading.

As any author will confess, writing a book is an absorbing – and at times mildly deranged – multi-year voyage. I could not have completed it without the patience, care and unfailing support of my family. To my wife, Crystal, and my sons, Orien and Shine: thank you for keeping me healthy, happy and – on most days – sane, and for helping me cross the finish line without tripping over my own enthusiasm. Crystal also offered sharp, thoughtful comments on a later draft of the manuscript – her first recorded approval of anything I've ever written. I take this as a promising omen for the book's success ... at least at home.

This book is meant to offer readers not only the latest scientific understanding of sex, but also a fresh perspective shaped by my own experiences and reflections as a research scientist. While I owe much to the many people who helped along the way, I alone am responsible for any errors that may have survived the journey.

A BRIEF GUIDE TO KEY TERMS

Anisogamy – Sexual reproduction in which the two gametes (sex cells) differ in size and often in form or function, with a large egg and a small sperm.

Autosome – Any chromosome that is not a sex chromosome. Humans usually have twenty-two pairs of autosomes (forty-four in total) and one pair of sex chromosomes (XX or XY). Autosomes carry most of the genes for traits like eye colour, height and metabolism.

Barr body – An inactivated X chromosome found in the cells of female mammals.

Bateman's rule – An evolutionary principle stating that male reproductive success usually increases with the number of mates, while female reproductive success is limited mainly by resources used for egg production. It's often shortened as females follow resources and males follow females.

Conjugation – In bacteria, the direct transfer of DNA from one cell to another through a connecting structure called a pilus. A form of horizontal gene transfer.

Cooperative bargaining – A negotiation process where parties work together to reach an agreement that benefits all sides, rather than maximizing only individual gain.

Cost of sex/cost of males – The evolutionary cost of sexual reproduction: in most species, only females bear offspring, while males mainly contribute sperm. This means that, compared to asexual reproduction, sexual populations may produce offspring at only about half the rate.

Diploid (2*n*) – Having two complete sets of chromosomes, one from each parent. In humans, most cells are diploid with forty-six chromosomes (twenty-three pairs).

Fisher's runaway hypothesis – The idea that a preferred trait (like a peacock's tail) and the preference for it can reinforce each other in a feedback loop, producing exaggerated traits over generations.

Gender – Socially: roles, behaviours and identities that societies associate with being male, female, both, neither, or somewhere in between. Biologically: the set of traits influencing reproduction in a given socioecological context. Gender may change over time and may not match the sex assigned at birth.

Gender reversal/sex-role reversal – When individuals of a species take on the behavioural and other biological traits typical of the opposite sex, without changing their biological sex.

Genetic recombination – The exchange of DNA between paired chromosomes during meiosis, producing new combinations of genes.

Genetic rescue – Introducing individuals from another population to increase the genetic diversity and health of a small or inbred population.

Genomic imprinting – A phenomenon in which a gene's effect depends on whether it was inherited from the mother or the father. One parental copy is silenced.

Handicap hypothesis – The idea that some costly traits signal quality because only the fittest individuals can afford them. Examples: a peacock's heavy tail or a deer's large antlers.

A Brief Guide to Key Terms

Haploid (*n*) – Having only one complete set of chromosomes. In humans, sperm and egg cells are haploid, each carrying twenty-three chromosomes.

Hermaphrodite – An organism with both male and female reproductive organs, able to produce both eggs and sperm. Common in earthworms, snails, some fish and many flowering plants. In humans, the term is outdated; 'intersex' is preferred.

Inbreeding depression – Reduced survival or fertility caused by mating between close relatives, which increases the chance of having harmful recessive diseases.

Intersex – A condition in which an individual's biological sex traits (chromosomes, hormones or genitalia) vary from typical male or female patterns. Fertility may be reduced.

Isogamy – Sexual reproduction in which gametes are the same size and appearance, though they may differ in mating type.

Kin selection – Helping relatives reproduce can increase your own genetic success, because you share genes with them. This combined success is called inclusive fitness.

K-strategists/r-strategists – Two ends of a life history strategy spectrum. **r-strategists**: comparatively many offspring, little parental investment, short lifespans; **K-strategists**: often fewer offspring, high parental investment, longer lifespans.

Lek – A gathering of males in a display area where they perform courtship behaviours to attract visiting females. Males provide no resources beyond sperm.

Life cycle – The sequence of stages an organism passes through from birth to reproduction to death. In sexual species, it alternates between haploid gametes and a diploid organism, which are termed as the two forms of existence in this book.

Master-switch gene – A gene that controls the activity of many other genes, directing the formation of major body structures. Example: *Sry* gene, which triggers male development in most mammals.

Mate choice – The selection of reproductive partners, often based on traits that signal health, fertility or good genes.

Mate competition – Rivalry between members of the same sex for access to mates. Often involves displays, fighting or resource defence.

Mating type – A genetic system that determines compatibility in isogamous species with similar gametes, such as fungi, algae and protozoa. It can be seen as proto-sex.

Meiosis – A special kind of cell division that produces gametes with half the number of chromosomes of the parental cell and shuffles genetic material to increase diversity.

Muller's ratchet – A phenomenon in which harmful mutations can accumulate over generations in asexual lineages because there is no recombination to eliminate them.

Parthenogenesis – Reproduction from an unfertilized egg. Offspring are usually clones of the mother. Seen in some plants, insects and more than eighty species of vertebrates.

Red Queen hypothesis – The idea that species must keep evolving just to survive because parasites, predators and competitors are also evolving.

Sensory exploitation hypothesis – Some mating signals evolve by taking advantage of pre-existing sensory biases in the receiver.

Sex chromosomes – Chromosomes that are linked to an individual's sex. In humans, XX = female, XY = male in most cases.

Sex determination – The process by which an organism develops as male, female or another sexual form. Can depend on chromosomes, temperature or other factors.

A Brief Guide to Key Terms

Sex ratio – The proportion of males to females in a population, measured at conception, birth or adulthood.

Sex reversal – An adaptive change from one biological sex to the other during an organism's lifetime.

Sexual conflict – When reproductive interests differ between males and females, leading to traits that benefit one sex but harm the other.

Sexy-son hypothesis – Females may choose attractive mates even if they give no care, because sons with those traits will attract more mates themselves.

Social cost of sex – Conflict of interest between genetic relatives caused by the lower relatedness as a result of sexual reproduction.

Sperm competition – Competition between sperm from different males to fertilize the same set of eggs.

Teleonomy – The appearance of purpose in biological traits or behaviours, shaped by natural selection rather than conscious intent. It is the opposite of teleology.

Transduction – Horizontal transfer of DNA mostly between bacteria via viruses (bacteriophages).

Transformation – Uptake of free DNA from the environment by a bacterium, incorporating it into its genome. It is also a form of horizontal gene transfer.

Triploid ($3n$) – Having three complete sets of chromosomes. Often infertile in animals; may reproduce asexually.

Trivers–Willard hypothesis – The idea that parents in good condition are more likely to produce sons, while those in poor condition are more likely to produce daughters.

Weber–Fechner law – The perceived change in a stimulus depends on its proportional change relative to the starting intensity, not on the absolute change.

NOTES

Chapter 1: A Ratchet, a Curse and a Story of Emergence – The Origin of Sex

1. Ryder, O.A., Thomas, S., Judson, J.M., Romanov, M.N., Dandekar, S., Papp, J.C., Sidak-Loftis, L.C. et al., 2021. Facultative parthenogenesis in California condors. *Journal of Heredity*, 112, pp. 569–74.
2. Maynard Smith, J., 1978. *The Evolution of Sex* (Cambridge: Cambridge University Press).
3. Williams, G.C., 1971. Introduction. In: G.C. Williams (ed.) *Group Selection* (New Brunswick, NJ: Aldine Transactions).
4. Zhang, S., Liu, Y., Ma, Y., Wang, H., Zhao, Y., Kuntner, M. and Li, D., 2022. Male spiders avoid sexual cannibalism with a catapult mechanism. *Current Biology*, 32(8), pp. R354–R355.
5. Carlson, E.A., 2009. *Hermann Joseph Muller (1890–1967)* (Washington, DC: National Academy of Sciences).
6. Pontecorvo, G., 1968. Hermann Joseph Muller, 1890–1967. *Biogr. Mems Fell. R. Soc.* 14, pp. 348–89.
7. Muller, H.J., 1964. The relation of recombination to mutational advance. *Mutation Research/Fundamental and Molecular Mechanisms of Mutagenesis*, 1(1), pp. 2–9.
8. Gabriel, W., Lynch, M. and Bürger, R., 1993. Muller's ratchet and mutational meltdowns. *Evolution*, 47(6), pp. 1744–57.
9. Hutchison III, C.A., Chuang, R.Y., Noskov, V.N., Assad-Garcia, N., Deerinck, T.J., Ellisman, M.H., Gill, J., Kannan, K., Karas, B.J., Ma, L. and Pelletier, J.F., 2016. Design and synthesis of a minimal bacterial genome. *Science*, 351(6280), p. aad6253.
10. Salmon, T.B., Evert, B.A., Song, B. and Doetsch, P.W., 2004. Biological consequences of oxidative stress-induced DNA damage in *Saccharomyces cerevisiae*. *Nucleic Acids Research*, 32(12), pp. 3712–23.

11. Moran, N.A., 1996. Accelerated evolution and Muller's ratchet in endosymbiotic bacteria. *Proceedings of the National Academy of Sciences*, 93(7), pp. 2873–8.
12. Martijn, J., Vosseberg, J., Guy, L., Offre, P. and Ettema, T.J., 2022. Phylogenetic affiliation of mitochondria with Alpha-II and Rickettsiales is an artefact. *Nature Ecology & Evolution*, 6(12), pp. 1829–31.
13. Mazel, D., 2006. Integrons: agents of bacterial evolution. *Nature Reviews Microbiology*, 4(8), pp. 608–20.
14. Mark Welch, D.B. and Meselson, M., 2000. Evidence for the evolution of bdelloid rotifers without sexual reproduction or genetic exchange. *Science*, 288(5469), pp. 1211–15.
15. Maynard Smith, J., 1986. Evolution: contemplating life without sex. *Nature*, 324(6095), pp. 300–301.
16. Gladyshev, E.A., Meselson, M. and Arkhipova, I.R., 2008. Massive horizontal gene transfer in bdelloid rotifers. *Science*, 320(5880), pp. 1210–13. Eyres, I., Boschetti, C., Crisp, A., Smith, T.P., Fontaneto, D., Tunnacliffe, A. and Barraclough, T.G., 2015. Horizontal gene transfer in bdelloid rotifers is ancient, ongoing and more frequent in species from desiccating habitats. *BMC Biology*, 13(1), pp. 1–17. Nowell, R.W., Rodriguez, F., Hecox-Lea, B.J., Mark Welch, D.B., Arkhipova, I.R., Barraclough, T.G. and Wilson, C.G., 2024. Bdelloid rotifers deploy horizontally acquired biosynthetic genes against a fungal pathogen. *Nature Communications*, 15(1), p. 5787.
17. Van Etten, J. and Bhattacharya, D., 2020. Horizontal gene transfer in eukaryotes: not if, but how much? *Trends in Genetics*, 36(12), pp. 915–25.
18. Ambur, O.H., Engelstädter, J., Johnsen, P.J., Miller, E.L. and Rozen, D.E., 2016. Steady at the wheel: conservative sex and the benefits of bacterial transformation. *Philosophical Transactions of the Royal Society B: Biological Sciences*, 371(1706), p. 20150528.
19. Otto, S.P., 2009. The evolutionary enigma of sex. *The American Naturalist*, 174(S1), pp. S1–S14.
20. Becks, L. and Agrawal, A.F., 2012. The evolution of sex is favoured during adaptation to new environments. *PLoS Biology*, 10(5), e1001317.
21. Neiman, M., Meirmans, S. and Meirmans, P.G., 2009. What can asexual lineage age tell us about the maintenance of sex? *Annals of the New York Academy of Sciences*, 1168(1), pp. 185–200. Speijer, D., Lukeš, J. and Eliáš, M., 2015. Sex is a ubiquitous, ancient, and inherent attribute of eukaryotic life. *Proceedings of the National Academy of Sciences*, 112(29), pp. 8827–34.

Notes

22 Jaron, K.S., Parker, D.J., Anselmetti, Y., Tran Van, P., Bast, J., Dumas, Z., Figuet, E., François, C.M., Hayward, K., Rossier, V. and Simion, P., 2022. Convergent consequences of parthenogenesis on stick insect genomes. *Science Advances*, 8(8), p. eabg3842.
23 Green, R.F. and Noakes, D.L.G., 1995. Is a little bit of sex as good as a lot? *Journal of Theoretical Biology*, 174(1), pp. 87–96. Hojsgaard, D. and Hörandl, E., 2015. A little bit of sex matters for genome evolution in asexual plants. *Frontiers in Plant Science*, 6, p. 82. Burke, N.W. and Bonduriansky, R., 2017. Sexual conflict, facultative asexuality, and the true paradox of sex. *Trends in Ecology & Evolution*, 32(9), pp. 646–52.
24 Baudry, E., Kryger, P., Allsopp, M., Koeniger, N., Vautrin, D., Mougel, F., Cornuet, J.-M. and Solignac, M., 2004. Whole-genome scan in thelytokous-laying workers of the Cape honeybee (*Apis mellifera capensis*): central fusion, reduced recombination rates and centromere mapping using half-tetrad analysis. *Genetics*, 167(1), pp. 243–52. Schwander, T. and Crespi, B.J., 2009. Multiple direct transitions from sexual reproduction to apomictic parthenogenesis in *Timema* stick insects. *Evolution*, 63(1), pp. 84–103. Archetti, M., 2010. Complementation, genetic conflict, and the evolution of sex and recombination. *Journal of Heredity*, 101(suppl_1), pp. S21–S33. Nougué, O., Rode, N.O., Jabbour-Zahab, R., Ségard, A., Chevin, L.-M., Haag, C.R. and Lenormand, T., 2015. Automixis in *Artemia*: solving a century-old controversy. *Journal of Evolutionary Biology*, 28(12), pp. 2337–48.
25 Griffiths, J.G. and Bonser, S.P., 2013. Is sex advantageous in adverse environments? A test of the abandon-ship hypothesis. *The American Naturalist*, 182(6), pp. 718–25.
26 Kokko, H., 2020. When synchrony makes the best of both worlds even better: How well do we really understand facultative sex? *The American Naturalist*, 195, pp. 380–92.
27 Agrawal, A.F., Hadany, L. and Otto, S.P., 2005. The evolution of plastic recombination. *Genetics*, 171, pp. 803–12.
28 Kleiven, O.T., Larsson, P. and Hobæk, A., 1992. Sexual reproduction in *Daphnia magna* requires three stimuli. *Oikos*, 63, pp. 197–206.
29 Barley, A.J., Reeder, T.W., Nieto-Montes de Oca, A., Cole, C.J. and Thomson, R.C., 2021. A new diploid parthenogenetic whiptail lizard from Sonora, Mexico, is the 'missing link' in the evolutionary transition to polyploidy. *The American Naturalist*, 198(2), pp. 295–309.
30 Lutes, A.A., Neaves, W.B., Baumann, D.P., Wiegraebe, W. and Baumann, P., 2010. Sister chromosome pairing maintains heterozygosity in parthenogenetic lizards. *Nature*, 464(7286), pp. 283–6.

31 Newton, A.A., Schnittker, R.R., Yu, Z., Munday, S.S., Baumann, D.P., Neaves, W.B. and Baumann, P., 2016. Widespread failure to complete meiosis does not impair fecundity in parthenogenetic whiptail lizards. *Development*, 143(23), pp. 4486–94.

32 Cox, M.M., 2001. Historical overview: searching for replication help in all of the rec places. *Proceedings of the National Academy of Sciences*, 98(15), pp. 8173–80. Bernstein, H. and Bernstein, C., 2019. Sexual processes in microbial Eukaryotes. In: *Parasitology and Microbiology Research*. IntechOpen, p. 135. Available at: www.intechopen.com/chapters/68651 [accessed 27 August 2025].

Chapter 2: Running with the Red Queen – *The Maintenance of Sex*

1 Nieuwenhuis, B.P.S. and James, T.Y., 2015. The frequency of sex in fungi. *Philosophical Transactions of the Royal Society B*, 371:20150540.

2 Kokko, H., 2020. When synchrony makes the best of both worlds even better: how well do we really understand facultative sex? *The American Naturalist*, 195(2), pp. 380–92.

3 Chen, L. and Wiens, J.J., 2021. Multicellularity and sex helped shape the Tree of Life. *Proceedings of the Royal Society B*, 288:20211265.

4 West, S.A., Lively, C.M. and Read, A.F., 1999. Sex may need more than one. *Journal of Evolutionary Biology*, 12, pp. 1053–5. Park, A.W., Jokela, J. and Michalakis, Y., 2010. Parasites and deleterious mutations: interactions influencing the evolutionary maintenance of sex. *Journal of Evolutionary Biology*, 23, pp. 1013–23.

5 University of Chicago, 2010. Leigh Van Valen: evolutionary theorist and paleobiology pioneer (1935–2010). Available at: https://news.uchicago.edu/story/leigh-van-valen-evolutionary-theorist-and-paleobiology-pioneer-1935-2010 [accessed 27 August 2025].

6 Van Valen, L., 1973. A new evolutionary law. *Evolutionary Theory*, 1, pp. 1–30.

7 Strotz, L.C., Simões, M., Girard, M.G., Breitkreuz, L., Kimmig, J. and Lieberman, B.S., 2018. Getting somewhere with the Red Queen: chasing a biologically modern definition of the hypothesis. *Biology Letters*, 14:20170734.

8 Lively, C.M. and Dybdahl, M.F., 2000. Parasite adaptation to locally common host genotypes. *Nature*, 405, pp. 679–81. Duncan, A. and Little, T.J., 2007. Parasite-driven genetic change in a natural population of *Daphnia*. *Evolution*, 64, pp. 796–803. Wolinska, J. and Spaak, P., 2009. The cost of being common: evidence from natural *Daphnia* populations. *Evolution*, 63, pp. 1893–901.

Notes

9. Lively, C.M., 2024. *Through the Looking Glass: I. Why Cross-Fertilize?* (Bloomington, IN: Indiana University Bloomington Libraries Publishing). Available at: doi.org/10.5967/GBD3-KA07 [accessed 27 August 2025].
10. Lively, C.M., 1987. Evidence from a New Zealand snail for the maintenance of sex by parasitism. *Nature*, 328, pp. 519–21.
11. Lively, C.M. and Jokela, J., 2002. Temporal and spatial distributions of parasites and sex in a freshwater snail. *Evolutionary Ecology Research*, 4, pp. 219–26.
12. Morran, L.T., Schmidt, O.G., Gelarden, I.A. II, Parrish, R.C. and Lively, C.M., 2011. Running with the Red Queen: host-parasite coevolution selects for biparental sex. *Science*, 333(6039), pp. 216–18.
13. Kerstes, N.A.G., Berenos, C., Schmid-Hempel, P. and Wegner, K.M., 2012. Antagonistic experimental coevolution with a parasite increases host recombination frequency. *BMC Evolutionary Biology*, 12:18.
14. Singh, N.D., Criscoe, D.R., Skolfield, S., Kohl, K.P., Keebaugh, E.S. and Schlenke, T.A., 2015. Fruit flies diversify their offspring in response to parasite infection. *Science*, 349, pp. 747–50.
15. Moritz, C., McCallum, H., Donnellan, S. and Roberts, J.D., 1991. Parasite loads in parthenogenetic and sexual lizards (*Heteronotia binoei*): support for the Red Queen hypothesis. *Proceedings of the Royal Society of London. Series B: Biological Sciences*, 244(1310), pp. 145–9.
16. Verhoeven, K.J.F. and Biere, A., 2013. Geographic parthenogenesis and plant-enemy interactions in the common dandelion. *BMC Evolutionary Biology*, 13:23.
17. Lively, C.M., Clark Craddock and Vrijenhoek, R.C., 1990. The Red Queen hypothesis supported by parasitism in sexual and clonal fish. *Nature*, 344, pp. 864–6.
18. Zhu, Y., Chen, H., Fan, J., Wang, Y., Li, Y., Chen, J. et al., 2000. Genetic diversity and disease control in rice. *Nature*, 406, pp. 718–22.
19. Charlesworth, D. and Willis, J.H., 2009. The genetics of inbreeding depression. *Nature Reviews Genetics*, 10, pp. 783–96.
20. Keller, L.F. and Waller, D.M., 2002. Inbreeding effects in wild populations. *Trends in Ecology & Evolution*, 17, pp. 230–41.
21. Westemeier, R.L. et al., 1998. Tracking the long-term decline and recovery of an isolated population. *Science*, 282, pp. 1695–8. Hedrick, P.W., 2001. Conservation genetics: where are we now? *Trends in Ecology & Evolution*, 16, pp. 629–36.
22. Hedrick, P.W. et al., 2014. Genetic rescue in Isle Royale wolves: genetic

analysis and the collapse of the population. *Conservation Genetics*, 15, pp. 1111–21.

23 Wilson, C.G. and Sherman, P.W., 2010. Anciently asexual bdelloid rotifers escape lethal fungal parasites by drying up and blowing away. *Science*, 327, pp. 547–76. Wilson, C.G. and Sherman, P.W., 2013. Spatial and temporal escape from fungal parasitism in natural communities of anciently asexual bdelloid rotifers. *Proceedings of the Royal Society of London. Series B: Biological Sciences*, 280:20131255.

24 Orellana, R., Macaya, C., Bravo, G., Dorochesi, F., Cumsille, A., Valencia, R., Rojas, C. and Seeger, M., 2018. Living at the frontiers of life: extremophiles in Chile and their potential for bioremediation. *Frontiers in Microbiology*, 9:2309.

25 Stephens, P.A., Sutherland, W.J. and Freckleton, R.P., 1999. What is the Allee effect? *Oikos*, 87(1), pp. 185–90. Courchamp, F., Berec, L. and Gascoigne, J., 2008. *Allee Effects in Ecology and Conservation* (Oxford: Oxford University Press).

26 Palomares, F., Godoy, J.A., López-Bao, J.V., Rodriguez, A., Roques, S., Casas-Marce, M., Revilla, E. and Delibes, M., 2012. Possible extinction vortex for a population of Iberian lynx on the verge of extirpation. *Conservation Biology*, 26(4), pp. 689–97.

27 Lively, C.M., in preparation. *Through the Looking Glass: II. Host-Parasite Coevolution and Sex*.

28 Dybdahl, M.F. and Lively, C.M., 1995. Host–parasite interactions: infection of common clones in natural populations of a freshwater snail (*Potamopyrgus antipodarum*). *Proceedings of the Royal Society of London. Series B: Biological Sciences*, 260(1357), pp. 99–103.

29 Vergara, D., Jokela, J. and Lively, C.M., 2014. Infection dynamics in coexisting sexual and asexual host populations: support for the Red Queen hypothesis. *The American Naturalist*, 184(S1), pp. S22–S30.

30 Burt, A. and Bell, G., 1987. Mammalian chiasma frequencies as a test of two theories of recombination. *Nature*, 326, pp. 803–5.

31 Wolfe, K.H. and Sharp, P.M., 1993. Mammalian gene evolution: nucleotide-sequence divergence between mouse and rat. *Journal of Molecular Evolution*, 37(4), pp. 441–56. Kuma, K., Iwabe, N. and Miyata, T., 1995. Functional constraints against variations on molecules from the tissue level: Slowly evolving brain-specific genes demonstrated by protein-kinase and immunoglobulin supergene families. *Molecular Biology and Evolution*, 12(1), pp. 123–30.

32 West, S.A., Gemmill, A.W., Graham, A., Viney, M.E. and Read, A.F.,

2001. Immune stress and facultative sex in a parasitic nematode. *Journal of Evolutionary Biology*, 14, pp. 333–7.
33. Vergara, D., Jokela, J., King, K.C. and Lively, C.M., 2013. The geographic mosaic of sex and infection in lake populations of a New Zealand snail at multiple spatial scales. *The American Naturalist*, 182(4), pp. 484–93.
34. Koskella, B. and Lively, C.M., 2009. Evidence for negative frequency-dependent selection during experimental coevolution of a freshwater snail and a sterilizing trematode. *Evolution*, 63(9), pp. 2213–21.
35. Haafke, J., Chakra, M.A. and Becks, L., 2016. Eco-evolutionary feedback promotes Red Queen dynamics and selects for sex in predator populations. *Evolution*, 70, pp. 641–52.
36. Gaba, S. and Ebert, D., 2009. Time-shift experiments as a tool to study antagonistic coevolution. *Trends in Ecology & Evolution*, 24, pp. 226–32. Svensson, E.I. and Råberg, L., 2010. Resistance and tolerance in animal enemy–victim coevolution. *Trends in Ecology & Evolution*, 25, pp. 267–74. Brockhurst, M.A., Chapman, T., King, K.C., Mank, J.E., Paterson, S. and Hurst, G.D.D., 2014. Running with the Red Queen: the role of biotic conflicts in evolution. *Proceedings of the Royal Society B*, 281:20141382. Liow, L.H., van Valen, L. and Stenseth, N.C., 2011. Red Queen: from populations to taxa and communities. *Trends in Ecology & Evolution*, 26, pp. 349–58. Strotz, L.C., Simões, M., Girard, M.G., Breitkreuz, L., Kimmig, J. and Lieberman, B.S., 2018. Getting somewhere with the Red Queen: chasing a biologically modern definition of the hypothesis. *Biology Letters*, 14:20170734.
37. Otto, S.P., 2009. The evolutionary enigma of sex. *The American Naturalist*, 174(S1), pp. S1–S14.

Chapter 3: Male, Female and Beyond – *Sex and Diversity*

1. Churchill, F.B., 2015. *August Weismann: Development, Heredity, and Evolution* (Cambridge, MA: Harvard University Press).
2. Nieuwenhuis, B.P. and Aanen, D.K., 2012. Sexual selection in fungi. *Journal of Evolutionary Biology*, 25, pp. 2397–411. Vreeburg, S., Nygren, K. and Aanen, D.K., 2016. Unholy marriages and eternal triangles: how competition in the mushroom life cycle can lead to genomic conflict. *Philosophical Transactions of the Royal Society B*, 371, 20150533.
3. Nieuwenhuis, B.P.S. and James, T.Y., 2015. The frequency of sex in fungi. *Philosophical Transactions of the Royal Society B*, 371, 20150540.

4 Lehtonen, J., Kokko, H. and Parker, G.A., 2016. What do isogamous organisms teach us about sex and the two sexes? *Philosophical Transactions of the Royal Society B*, 371, 20150532.
5 Ibid.; Bachtrog, D. et al., 2014. Sex determination: why so many ways of doing it? *PLoS Biology*, 12, e1001899.
6 Parker, G.A., Baker, R.R. and Smith, V.G.F., 1972. The origin and evolution of gamete dimorphism and the male-female phenomenon. *Journal of Theoretical Biology*, 36(3), pp. 529–53.
7 Dyer, P.S. and Kück, U., 2017. Sex and the imperfect fungi. *Microbiology Spectrum*, 5(3), FUNK-0043-2017.
8 Czárán, T.L. and Hoekstra, R.F., 2004. Evolution of sexual asymmetry. *BMC Evolutionary Biology*, 4, pp. 1–12. Hadjivasiliou, Z. and Pomiankowski, A., 2016. Gamete signalling underlies the evolution of mating types and their number. *Philosophical Transactions of the Royal Society B*, 371, 20150531.
9 Ni, M., Feretzaki, M., Sun, S., Wang, X. and Heitman, J., 2011. Sex in fungi. *Annual Review of Genetics*, 45, pp. 405–30. Freihorst, D., Fowler, T.J., Bartholomew, K., Raudaskoski, M., Horton, J.S. and Kothe, E., 2016. The mating-type genes of basidiomycetes. In: J. Wendland (ed.), *The Mycota I: Growth, Differentiation and Sexuality*, 3rd edn (Basel: Springer), pp. 329–49. Vreeburg, S., Nygren, K. and Aanen, D.K., 2016. Unholy marriages and eternal triangles: how competition in the mushroom life cycle can lead to genomic conflict. *Philosophical Transactions of the Royal Society B*, 371, 20150533.
10 Kothe, E., 1999. Mating types and pheromone recognition in the homobasidiomycete *Schizophyllum commune*. *Fungal Genetics and Biology*, 27, pp. 146–52.
11 Hadjivasiliou, Z., Iwasa, Y. and Pomiankowski, A., 2015. Cell–cell signalling in sexual chemotaxis: a basis for gametic differentiation, mating types and sexes. *Journal of the Royal Society Interface*, 12, p. 20150342. Krumbeck, Y., Constable, G.W.A. and Rogers, T., 2020. Fitness differences suppress the number of mating types in evolving isogamous species. *Royal Society Open Science*, 7, p. 192126.
12 Iwasa, Y. and Sasaki, A., 1987. Evolution of the number of sexes. *Evolution*, 41, pp. 49–65. Billiard, S., López-Villavicencio, M., Devier, B., Hood, M.E., Fairhead, C. and Giraud, T., 2011. Having sex, yes, but with whom? Inferences from fungi on the evolution of anisogamy and mating types. *Biological Reviews*, 86, pp. 421–42. Hadjivasiliou, Z., Iwasa, Y. and Pomiankowski, A., 2015. Cell–cell signalling in sexual

chemotaxis: a basis for gametic differentiation, mating types and sexes. *Journal of the Royal Society Interface*, 12, p. 20150342.

13 Constable, G.W.A. and Kokko, H., 2018. The rate of facultative sex governs the number of expected mating types in isogamous species. *Nature Ecology & Evolution*, 2, pp. 1168–75. Krumbeck, Y., Constable, G.W.A. and Rogers, T., 2020. Fitness differences suppress the number of mating types in evolving isogamous species. *Royal Society Open Science*, 7, p. 192126.

14 Lehtonen, J., Kokko, H. and Parker, G.A., 2016. What do isogamous organisms teach us about sex and the two sexes? *Philosophical Transactions of the Royal Society B: Biological Sciences*, 371(1706), p. 20150532.

15 Ibid.; Togashi, T., Horinouchi, Y. and Parker, G.A., 2021. A comparative test of the gamete dynamics theory for the evolution of anisogamy in *Bryopsidales* green algae. *Royal Society Open Science*, 8(3), p. 201611.

16 Parker, G.A., Baker, R.R. and Smith, V.G.F., 1972. The origin and evolution of gamete dimorphism and the male–female phenomenon. *Journal of Theoretical Biology*, 36(3), pp. 529–53. Lehtonen, J. and Kokko, H., 2011. Two roads to two sexes: Unifying gamete competition and gamete limitation in a single model of anisogamy evolution. *Behavioral Ecology and Sociobiology*, 65, pp. 445–59. Togashi, T., Bartelt, J.L., Yoshimura, J., Tainaka, K. and Cox, P.A., 2012. Evolutionary trajectories explain the diversified evolution of isogamy and anisogamy in marine green algae. *Proceedings of the National Academy of Sciences USA*, 109, pp. 13692–7.

17 Karp, N.A. et al., 2017. Prevalence of sexual dimorphism in mammalian phenotypic traits. *Nature Communications*, 8(1), p. 15475.

18 Winkler, L. and Lindholm, A.K., 2022. A meiotic driver alters sperm form and function in house mice: a possible example of spite. *Chromosome Research*, 30(2–3), pp. 151–64.

19 Haig, D., 2020. *From Darwin to Derrida: Selfish Genes, Social Selves, and the Meanings of Life* (Cambridge, MA: MIT Press).

20 Cruz, R., Carballo, M., Conde-Padín, P. and Rolán-Alvarez, E., 2004. Testing alternative models for sexual isolation in natural populations of *Littorina saxatilis*: Indirect support for by-product ecological speciation? *Journal of Evolutionary Biology*, 17(2), pp. 288–93. Chen, L. and Wiens, J.J., 2021. Multicellularity and sex helped shape the Tree of Life. *Proceedings of the Royal Society B*, 288(1955), p. 20211265.

21 Bokma, F. et al., 2016. Testing for Depéret's rule (body size increase) in mammals using combined extinct and extant data. *Systematic Biology*, 65(1), pp. 98–108.

22 Bush, A.M., Hunt, G. and Bambach, R.K., 2016. Sex and the shifting biodiversity dynamics of marine animals in deep time. *Proceedings of the National Academy of Sciences*, 113(49), pp. 14073–8.
23 Chen, L. and Wiens, J.J., 2021. Multicellularity and sex helped shape the Tree of Life. *Proceedings of the Royal Society B*, 288(1955), p. 20211265.

Chapter 4: The Makings and Variations of Sex –
Sex Determination and Development

1 Ryder, O.A., Thomas, S., Judson, J.M., Romanov, M.N., Dandekar, S., Papp, J.C., Sidak-Loftis, L.C. et al., 2021. Facultative parthenogenesis in California condors. *Journal of Heredity*, 112(7), pp. 569–74.
2 Grützner, F., Rens, W., Tsend-Ayush, E., El-Mogharbel, N., O'Brien, P.C., Jones, R.C., Ferguson-Smith, M.A. and Marshall Graves, J.A., 2004. In the platypus a meiotic chain of ten sex chromosomes shares genes with the bird Z and mammal X chromosomes. *Nature*, 432(7019), pp. 913–17.
3 Gempe, T., Hasselmann, M., Schiøtt, M., Hause, G., Otte, M. and Beye, M., 2009. Sex determination in honeybees: two separate mechanisms induce and maintain the female pathway. *PLoS Biology*, 7(10), p. e1000222.
4 Traynor, K.S., Le Conte, Y. and Page, R.E., Jr, 2014. Queen and young larval pheromones impact nursing and reproductive physiology of honey bee (*Apis mellifera*) workers. *Behavioral Ecology and Sociobiology*, 68, pp. 2059–73.
5 Schultheiss, P., Nooten, S.S., Wang, R., Wong, M.K., Brassard, F. and Guénard, B., 2022. The abundance, biomass, and distribution of ants on Earth. *Proceedings of the National Academy of Sciences*, 119(40), p. e2201550119.
6 Lockley, E.C. and Eizaguirre, C., 2021. Effects of global warming on species with temperature-dependent sex determination: bridging the gap between empirical research and management. *Evolutionary Applications*, 14(10), pp. 2361–77.
7 Ge, C. et al., 2018. The histone demethylase KDM6B regulates temperature-dependent sex determination in a turtle species. *Science*, 360, pp. 645–8.
8 Bull, J.J., 1983. *Evolution of Sex Determining Mechanisms* (Menlo Park, CA: Benjamin Cummings).
9 Pen, I., Uller, T., Feldmeyer, B., Harts, A., While, G.M. et al., 2010. Climate-driven population divergence in sex-determining systems. *Nature*, 468, pp. 436–8.

Notes

10   Bull, J.J., 1983. *Evolution of Sex Determining Mechanisms* (Menlo Park, CA: Benjamin Cummings). Berec, L. et al., 2005. Sex determination in *Bonellia viridis* (Echiura: Bonelliidae): population dynamics and evolution. *OIKOS*, 108, pp. 473–84.

11   Jaccarini, V., Agius, L., Schembri, P.J. et al., 1983. Sex determination and larval sexual interaction in *Bonellia viridis* Rolando (Echiura: Bonelliidae). *Journal of Experimental Marine Biology and Ecology*, 66, pp. 25–40.

12   Bull, J.J., 1983. *Evolution of Sex Determining Mechanisms* (Menlo Park, CA: Benjamin Cummings). Kobayashi, Y., Nagahama, Y. and Nakamura, M., 2013. Diversity and plasticity of sex determination and differentiation in fishes. *Sex Development*, 7, pp. 115–25. Warner, R.R., Fitch, D.L. and Standish, J.D., 1996. Social control of sex change in the shelf limpet, *Crepidula norrisiarum*: size-specific responses to local group composition. *Journal of Experimental Marine Biology and Ecology*, 204, pp. 155–67.

13   Cheng, H., Guo, Y., Yu, Q. and Zhou, R., 2003. The rice field eel as a model system for vertebrate sexual development. *Cytogenetic and Genome Research*, 101(3–4), pp. 274–7.

14   Kobayashi, Y. et al., 2009. Sex change in the Gobiid fish is mediated through rapid switching of gonadotropin receptors from ovarian to testicular portion or vice versa. *Endocrinology*, 150, pp. 1503–11. Casadevall, M., Delgado, E., Colleye, O., Monserrat, S.B. and Parmentier, E., 2009. Histological study of the sex-change in the skunk clownfish *Amphiprion akallopisos*. *Open Fish Science Journal*, 2, pp. 55–8. Casas, L. et al., 2016. Sex change in clownfish: molecular insights from transcriptome analysis. *Scientific Reports*, 6, p. 35461.

15   Munday, P.L., Buston, P.M. and Warner, R.R., 2006. Diversity and flexibility of sex-change strategies in animals. *Trends in Ecology & Evolution*, 21, pp. 89–95.

16   Verhulst, E.C., Beukeboom, L.W. and van de Zande, L., 2010. Maternal control of haplodiploid sex determination in the wasp *Nasonia*. *Science*, 328, pp. 620–23. Beye, M., Hasselmann, M., Fondrk, M.K., Page, R.E. and Omholt, S.W., 2003. The gene *csd* is the primary signal for sexual development in the honeybee and encodes an SR-type protein. *Cell*, 114, pp. 419–29. Hediger, M., Henggeler, C., Meier, N., Perez, R., Saccone, G. et al., 2010. Molecular characterization of the key switch F provides a basis for understanding the rapid divergence of the sex-determining pathway in the housefly. *Genetics*, 184, pp. 155–70.

17   Nanda, I., Kondo, M., Hornung, U., Asakawa, S., Winkler, C. et al.,

2002. A duplicated copy of *DMRT1* in the sex-determining region of the Y chromosome of the medaka, *Oryzias latipes*. *Proceedings of the National Academy of Sciences of the United States of America*, 99, pp. 11778–83. Yoshimoto, S., Okada, E., Umemoto, H., Tamura, K., Uno, Y. et al., 2008. A W-linked DM-domain gene, *DM-W*, participates in primary ovary development in *Xenopus laevis*. *Proceedings of the National Academy of Sciences of the United States of America*, 105, pp. 2469–74. Smith, C.A., Roeszler, K.N., Ohnesorg, T., Cummins, D.M., Farlie, P.G. et al., 2009. The avian Z-linked gene *DMRT1* is required for male sex determination in the chicken. *Nature*, 461, pp. 267–71.

18 Haag, E.S. and Doty, A.V., 2005. Sex determination across evolution: connecting the dots. *PLoS Biology*, 3, p. e21. Kopp, A., 2012. Dmrt genes in the development and evolution of sexual dimorphism. *Trends in Genetics*, 28, pp. 175–84.

19 Navajas-Perez, R., Schwarzacher, T., Rejon, M.R., Garrido-Ramos, M.A., 2009. Molecular cytogenetic characterization of *Rumex papillaris*, a dioecious plant with an XX/XY(1)Y(2) sex chromosome system. *Genetica*, 135, pp. 87–93. Vicoso, B., Emerson, J.J., Zektser, Y., Mahajan, S., Bachtrog, D., 2013. Comparative sex chromosome genomics in snakes: differentiation, evolutionary strata, and lack of global dosage compensation. *PLoS Biology*, 11, p. e1001643. Carvalho, A.B., Koerich, L.B., Clark, A.G., 2009. Origin and evolution of Y chromosomes: *Drosophila* tales. *Trends in Genetics*, 25, pp. 270–77. Nanda, I., Schlegelmilch, K., Haaf, Y., Schartl, M., Schmid, M., 2008. Synteny conservation of the Z chromosome in 14 avian species (11 families) supports a role for Z dosage in avian sex determination. *Cytogenetic and Genome Research*, 122, pp. 150–56. Skaletsky, H., Kuroda-Kawaguchi, T., Minx, P., Cordum, H., Hillier, L. et al., 2003. The male-specific region of the human Y chromosome is a mosaic of discrete sequence classes. *Nature*, 423, pp. 825–37.

20 Graves, J., 2006. Sex chromosome specialization and degeneration in mammals. *Cell*, 124, pp. 901–14. Parma, P., Veyrunes, F. and Pailhoux, E., 2016. Sex reversal in non-human placental mammals. *Sexual Development*, 10(5–6), pp. 326–44.

21 Délot, E.C. and Vilain, E., 2021. Towards improved genetic diagnosis of human differences of sex development. *Nature Reviews Genetics*, 22(9), pp. 588–602.

22 Shearwin-Whyatt, L., Fenelon, J., Yu, H., Major, A., Qu, Z., Zhou, Y., Shearwin, K., Galbraith, J., Stuart, A., Adelson, D. and Zhang, G.,

2025. *AMHY* and sex determination in egg-laying mammals. *Genome Biology*, 26(1), p. 144.
23   Eggers, S. and Sinclair, A., 2012. Mammalian sex determination – insights from humans and mice. *Chromosome Research*, 20, pp. 215–38.
24   Von Stackelberg, K.T., 2020. hermaphroditism. In *Oxford Research Encyclopedia of Classics*. Available at: oxfordre.com/classics/display/10.1093/acrefore/9780199381135.001.0001/acrefore-9780199381135-e-8090.
25   Sax, L., 2002. How common is intersex? A response to Anne Fausto-Sterling. *Journal of Sex Research*, 39, pp. 174–8. Hughes, I.A., Houk, C., Ahmed, S.F. and Lee, P.A., 2006. Consensus statement on management of intersex disorders. *Journal of Pediatric Urology*, 2, pp. 148–62.
26   Mazen, I., Hiort, O., Bassiouny, R., El Gammal, M., 2008. Differential diagnosis of disorders of sex development in Egypt. *Hormone Research*, 70, pp. 118–23. Bashamboo, A. and McElreavey, K., 2016. Mechanisms of sex determination in humans: insights from disorders of sex development. *Sex Development*, 10, pp. 313–25.
27   Quinn, A. and Koopman, P., 2012. The molecular genetics of sex determination and sex reversal in mammals. *Seminars in Reproductive Medicine*, 30(5), pp. 351–63.
28   Hall, C.A., Conroy, G., Jelocnik, M., Kasimov, V., Gillet, A., Portas, T., Hill, A. and Potvin, D.A., 2025. Prevalence and implications of sex reversal in free-living birds. *Biology Letters*, 21(8), p. 20250182.
29   Chau, P.L. and Herring, J., 2021. *Emergent Medicine and the Law* (Cham: Palgrave Macmillan).
30   Fausto-Sterling, A., 2020. *Sexing the Body: Gender Politics and the Construction of Sexuality* (New York: Basic Books.
31   Prum, R.O., 2023. Performance all the way down: genes, development, and sexual difference. In: *Performance All the Way Down* (Chicago: University of Chicago Press).
32   Ellison, J., 2012. Caster Semenya and the IOC's Olympics gender bender. *The Daily Beast*. Available at: www.thedailybeast.com/caster-semenya-and-the-iocs-olympics-gender-bender/ [accessed 27 August 2025].
33   Xavier, N.A. and McGill, J.B., 2012. Hyperandrogenism and intersex controversies in women's Olympics. *Journal of Clinical Endocrinology & Metabolism*, 97, pp. 3902–7.
34   Savage, R., 2020. Olympic champion Caster Semenya's 11-year battle to compete. *Reuters*, 9 September.

Chapter 5: The Gender Bender – *Gender and Adaptation*

1. Emlen, S.T. and Wrege, P.H., Size dimorphism, intrasexual competition, and sexual selection in Wattled Jacana (*Jacana jacana*), a sex-role-reversed shorebird in Panama. *The Auk*, 121, pp. 391–403. Emlen, S.T. and Wrege, P.H., Division of labour in parental care behaviour of a sex-role-reversed shorebird, the wattled jacana. *Animal Behaviour*, 68, pp. 847–55.
2. Vincent, A.C.J. and Sadler, L.M., 1995. Faithful pair bonds in wild seahorses, *Hippocampus whitei*. *Animal Behaviour*, 50, pp. 1557–69.
3. United Nations, n.d. *Concepts and definitions*. [online] Available at: www.un.org/womenwatch/osagi/conceptsandefinitions.htm [accessed 25 March 2025].
4. Fuentes, A., 2025. *Sex Is a Spectrum: The Biological Limits of the Binary* (Princeton, NJ: Princeton University Press). DuBois, L.Z., Ritz, S.A., McCarthy, M.M. and Kaiser Trujillo, A. (eds), 2025. Sex and gender. *Sex and Gender* (Cham: Springer), pp. 1–21.
5. Yoshizawa, K., Ferreira, R.L., Kamimura, Y. and Lienhard, C., 2014. Female penis, male vagina, and their correlated evolution in a cave insect. *Current Biology*, 24(9), pp. 1006–10.
6. Yoshizawa, K., Ferreira, R.L., Kamimura, Y. and Lienhard, C., 2019. Why did a female penis evolve in a small group of cave insects? *BioEssays*, 41(6). Kamimura, Y., Yoshizawa, K., Lienhard, C., Ferreira, R.L. and Abe, J., 2021. Evolution of nuptial gifts and its coevolutionary dynamics with male-like persistence traits of females for multiple mating. *BMC Ecology and Evolution*, 21(1), p. 164.
7. Frankand, L.G. and Glickman, S.E., 1994. Giving birth through a penile clitoris: parturition and dystocia in the spotted hyaena (*Crocuta crocuta*). *Journal of Zoology*, 234(4), pp. 659–65.
8. Smith, J.E. and Holekamp, K.E., 2019. Spotted hyenas. In: J.C. Choe (ed.), *Encyclopedia of Animal Behavior*, 2nd edn (Oxford: Academic Press), pp. 190–208.
9. Guillette Jr, L.J., Gross, T.S., Masson, G.R., Matter, J.M., Percival, H.F. and Woodward, A.R., 1994. Developmental abnormalities of the gonad and abnormal sex hormone concentrations in juvenile alligators from contaminated and control lakes in Florida. *Environmental Health Perspectives*, 102(8), pp. 680–88.
10. Clark, M.M. and Galef Jr, B.G., 1990. Sexual segregation in the left and right horns of the gerbil uterus: 'The male embryo is usually on the right, the female on the left' (Hippocrates). *Developmental Psychobiology*, 23(1), pp. 29–37.

Notes

11 Clark, M.M., Crews, D. and Galef Jr, B.G., 1991. Concentrations of sex steroid hormones in pregnant and fetal Mongolian gerbils. *Physiology & Behavior*, 49(2), pp. 239–43. Clark, M.M., Vom Saal, F.S. and Galef Jr, B.G., 1992. Intrauterine positions and testosterone levels of adult male gerbils are correlated. *Physiology & Behavior*, 51(5), pp. 957–60.
12 Clark, M.M., Johnson, J. and Galef Jr, B.G., 2004. Sexual motivation suppresses paternal behaviour of male gerbils during their mates' postpartum oestrus. *Animal Behaviour*, 67(1), pp. 49–57.
13 Clark, M.M. and Galef Jr, B.G., 1995. Prenatal influences on reproductive life history strategies. *Trends in Ecology & Evolution*, 10(4), pp. 151–3.
14 Lummaa, V., Pettay, J.E. and Russell, A.F., 2007. Male twins reduce fitness of female co-twins in humans. *Proceedings of the National Academy of Sciences*, 104(26), pp. 10915–20.
15 Bütikofer, A., Figlio, D.N., Karbownik, K., Kuzawa, C.W. and Salvanes, K.G., 2019. Evidence that prenatal testosterone transfer from male twins reduces the fertility and socioeconomic success of their female co-twins. *Proceedings of the National Academy of Sciences*, 116(14), pp. 6749–53.
16 Nasoori, A., 2020. Formation, structure, and function of extra-skeletal bones in mammals. *Biological Reviews*, 95(4), pp. 986–1019.
17 Barrionuevo, F.J., Zurita, F., Burgos, M. and Jiménez, R., 2004. Testis-like development of gonads in female moles: new insights on mammalian gonad organogenesis. *Developmental Biology*, 268(1), pp. 39–52.
18 Zheng, Z. and Cohn, M.J., 2011. Developmental basis of sexually dimorphic digit ratios. *Proceedings of the National Academy of Sciences of the United States of America*, 108(39), pp. 16289–94.
19 Watts, T.M., Holmes, L., Raines, J., Orbell, S. and Rieger, G., 2018. Finger length ratios of identical twins with discordant sexual orientations. *Archives of Sexual Behavior*, 47(8), pp. 2435–44.
20 Sorokowski, P. and Kowal, M., 2024. Relationship between the 2D:4D and prenatal testosterone, adult level testosterone, and testosterone change: Meta-analysis of 54 studies. *American Journal of Biological Anthropology*, 183(1), pp. 20–38.
21 Bailey, J.M., Vasey, P.L., Diamond, L.M., Breedlove, S.M., Vilain, E. and Epprecht, M., 2016. Sexual orientation, controversy, and science. *Psychological Science in the Public Interest*, 17(2), pp. 45–101.
22 Hamer, D.H., Hu, S., Magnuson, V.L., Hu, N. and Pattatucci, A.M.L., 1993. A linkage between DNA markers on the X chromosome

and male sexual orientation. *Science*, 261, pp. 321–7. Sanders, A.R., Beecham, G.W., Guo, S., Dawood, K., Rieger, G., Badner, J.A., Gershon, E.S. et al., 2017. Genome-wide association study of male sexual orientation. *Scientific Reports*, 7(1), pp. 1–6.

23 Ganna, A., Verweij, K.J.H., Nivard, M.G., Maier, R., Wedow, R., Busch, A.S., Abdellaoui, A., Guo, S., Sathirapongsasuti, J.F., 23andMe Research Team and Lichtenstein, P., 2019. Large-scale GWAS reveals insights into the genetic architecture of same-sex sexual behavior. *Science*, 365(6456), p. eaat7693.

24 Borsa, A., Miyagi, M., Ichikawa, K., Jesus, K.D., Jillson, K., Boulicault, M. and Richardson, S.S., 2024. The new genetics of sexuality. *GLQ*, 30(1), pp. 119–40.

25 Cantar, J.M., Blanchard, R., Paterson, A.D. and Bogaert, A.F., 2002. How many gay men owe their sexual orientation to fraternal birth order? *Archives of Sexual Behavior*, 31, pp. 63–71. VanderLaan, D.P. and Vasey, P.L., 2011. Male sexual orientation in Independent Samoa: evidence for fraternal birth order and maternal fecundity effects. *Archives of Sexual Behavior*, 40, pp. 495–503.

26 Bailey, J.M., Vasey, P.L., Diamond, L.M., Breedlove, S.M., Vilain, E. and Epprecht, M., 2016. Sexual orientation, controversy, and science. *Psychological Science in the Public Interest*, 17(2), pp. 45–101.

27 Bogaert, A.F., Skorska, M.N., Wang, C., Gabrie, J., MacNeil, A.J., Hoffarth, M.R., VanderLaan, D.P., Zucker, K.J. and Blanchard, R., 2018. Male homosexuality and maternal immune responsivity to the Y-linked protein NLGN4Y. *Proceedings of the National Academy of Sciences*, 115(2), pp. 302–6.

28 Ibid.

29 Blanchard, R., 2018. Fraternal birth order, family size, and male homosexuality: Meta-analysis of studies spanning 25 years. *Archives of Sexual Behavior*, 47(1), pp. 1–15. Flegr, J. and Kuba, R., 2024. Birth order: parental manipulation hypothesis. In: T.K. Shackelford and V.A. Weekes-Shackelford (eds), *Encyclopedia of Sexual Psychology and Behavior* (Cham: Springer), pp. 1–6.

30 Bhattacharya, S., Amodei, R., Vilain, E. and Roselli, C.E., 2022. Identification of differential hypothalamic DNA methylation and gene expression associated with sexual partner preferences in rams. *PLoS ONE*, 17(5), p. e0263319.

31 Nugent, B.M., Wright, C.L., Shetty, A.C., Hodes, G.E., Lenz, K.M., Mahurkar, A. et al., 2015. Brain feminization requires active repression

of masculinization via DNA methylation. *Nature Neuroscience*, 18, pp. 690–97. Cisternas, C.D., Cortes, L.R., Golynker, I., Castillo-Ruiz, A. and Forger, N.G., 2020. Neonatal inhibition of DNA methylation disrupts testosterone-dependent masculinization of neurochemical phenotype. *Endocrinology*, 161(1).

32 Ryan, B.C. and Vandenbergh, J.G., 2002. Intrauterine position effects. *Neuroscience & Biobehavioral Reviews*, 26(6), pp. 665–78.

33 Roselli, C.E., Stadelman, H., Reeve, R., Bishop, C.V. and Stormshak, F., 2007. The ovine sexually dimorphic nucleus of the medial preoptic area is organized prenatally by testosterone. *Endocrinology*, 148(9), pp. 4450–57.

34 Bhattacharya, S., Amodei, R., Vilain, E. and Roselli, C.E., 2022. Identification of differential hypothalamic DNA methylation and gene expression associated with sexual partner preferences in rams. *PLoS ONE*, 17(5), p. e0263319.

35 MacFarlane, G.R., Blomberg, S.P. and Vasey, P.L., 2010. Homosexual behaviour in birds: frequency of expression is related to parental care disparity between the sexes. *Animal Behaviour*, 80(3), pp. 375–90. Monk, J.D., Giglio, E., Kamath, A., Lambert, M.R. and McDonough, C.E., 2019. An alternative hypothesis for the evolution of same-sex sexual behaviour in animals. *Nature Ecology & Evolution*, 3(12), pp. 1622–31. Gómez, J.M., González-Megías, A. and Verdú, M., 2023. The evolution of same-sex sexual behaviour in mammals. *Nature Communications*, 14(1), p. 5719. Anderson, K.A., Teichroeb, J.A., Ramsay, M.S., Bădescu, I., López-Torres, S. and Gibb, J.K., 2024. Same-sex sexual behaviour among mammals is widely observed, yet seldomly reported: Evidence from an online expert survey. *PLoS ONE*, 19(6), p. e0304885.

36 Ganna, A., Verweij, K.J.H., Nivard, M.G., Maier, R., Wedow, R., Busch, A.S., Abdellaoui, A., Guo, S., Sathirapongsasuti, J.F., 23andMe Research Team and Lichtenstein, P., 2019. Large-scale GWAS reveals insights into the genetic architecture of same-sex sexual behavior. *Science*, 365(6456), p. eaat7693.

37 Gómez, J.M., González-Megías, A. and Verdú, M., 2023. The evolution of same-sex sexual behaviour in mammals. *Nature Communications*, 14(1), p. 5719. Lents, N.H., 2025. *The Sexual Evolution: How 500 Million Years of Sex, Gender, and Mating Shape Modern Relationships* (New York: HarperCollins).

38 Jahme, C., 2000. *Beauty and the Beasts: Woman, Ape and Evolution* (London: Virago).

39. Song, S. and Zhang, J., 2024. Genetic variants underlying human bisexual behavior are reproductively advantageous. *Science Advances*, 10, p. eadj6958.
40. Von Rueden, C.R. and Jaeggi, A.V., 2016. Men's status and reproductive success in 33 nonindustrial societies: Effects of subsistence, marriage system, and reproductive strategy. *Proceedings of the National Academy of Sciences*, 113, pp. 10824–9.
41. Zietsch, B.P., Sidari, M.J., Abdellaoui, A., Maier, R., Langström, N., Guo, S., Beecham, G.W., Martin, E.R., Sanders, A.R. and Verweij, K.J.H., 2021. Genomic evidence consistent with antagonistic pleiotropy may help explain the evolutionary maintenance of same-sex sexual behaviour in humans. *Nature Human Behaviour*, 5, pp. 1251–8.
42. Foster, S.J. and Vincent, A.C.J., 2004. Life history and ecology of seahorses: implications for conservation and management. *Journal of Fish Biology*, 65(1), pp. 1–61.
43. Cassar, A. and Zhang, Y.J., 2022. The competitive woman: evolutionary insights and cross-cultural evidence into finding the *Femina Economica*. *Journal of Economic Behavior & Organization*, 197, pp. 447–71.
44. Sun, L., 2023. *The Liars of Nature and the Nature of Liars: Cheating and Deception in the Living World* (Princeton, NJ: Princeton University Press).
45. Hampson, E. and Kimura, D.K., 1988. Reciprocal effects of hormonal fluctuations on human motor and perceptual-spatial skills. *Behavioral Neuroscience*, 102, pp. 456–9.
46. Wikipedia, n.d. Same-sex marriage legislation in the United States. [online] Available at: en.wikipedia.org/wiki/Same-sex_marriage_legislation_in_the_United_States [accessed 7 June 2025].

Chapter 6: Red in Tooth and Claw – *Mate Competition*

1. Alexander, C. (2003). *The Bounty: The True Story of the Mutiny of the Bounty* (New York: Viking).
2. Aristotle, *Politics*.
3. Lindenfors, P., Gittleman, J.L. and Jones, K.E., 2007. Sexual size dimorphism in mammals. In: Fairbairn, D.J., Blanckenhorn, W.U. and Szekely, T. (eds), *Sex, Size and Gender Roles: Evolutionary Studies of Sexual Size Dimorphism* (Oxford: Oxford University Press), pp. 19–26. However, not all researchers agree. See Bro-Jørgensen, J., 2007. The intensity of sexual selection predicts weapon size in male bovids. *Evolution*, 61, pp. 1316–26.

Notes

4   Kopp, A., 2012. Dmrt genes in the development and evolution of sexual dimorphism. *Trends in Genetics*, 28(4), pp. 175–84.
5   Blanckenhorn, W.U., 2005. Behavioral causes and consequences of sexual size dimorphism. *Ethology*, 111(11), pp. 977–1016.
6   Doody, J.S., Dinets, V. and Burghardt, G.M., 2021. *The Secret Social Lives of Reptiles* (Baltimore, MD: Johns Hopkins University Press).
7   Clutton-Brock, T.H., Albon, S.D., Gibson, R.M. and Guinness, F.E., 1979. The logical stag: adaptive aspects of fighting in red deer (*Cervus elaphus* L.). *Animal Behaviour*, 27, pp. 211–25.
8   Fisher, H.S. et al., 2014. The dynamics of sperm cooperation in a competitive environment. *Proceedings of the Royal Society B: Biological Sciences*, 281(1790), p. 20140296. Fisher, H.S. et al., 2016. The genetic basis and fitness consequences of sperm midpiece size in deer mice. *Nature Communications*, 7, p. 13652.
9   Civetta, A. and Ranz, J.M., 2019. Genetic factors influencing sperm competition. *Frontiers in Genetics*, 10, p. 820.
10 Waage, J.K., 1979. Dual function of the damselfly penis: sperm removal and transfer. *Science*, 203(4383), pp. 916–18.
11 Davies, N.B., 1983. Polyandry, cloaca-pecking and sperm competition in dunnocks. *Nature*, 302(5906), pp. 334–6.
12 Shine, R., Olsson, M.M. and Mason, R.T., 2000. Chastity belts in gartersnakes: the functional significance of mating plugs. *Biological Journal of the Linnean Society*, 70(3), pp. 377–90.
13 Baker, R.R. and Shackelford, T.K., 2018. A comparison of paternity data and relative testes size as measures of level of sperm competition in the Hominoidea. *American Journal of Physical Anthropology*, 165(3), pp. 421–43.
14 Vaughn, A.A., delBarco-Trillo, J. and Ferkin, M.H., 2008. Sperm investment in male meadow voles is affected by the condition of the nearby male conspecifics. *Behavioral Ecology*, 19(6), pp. 1159–64.
15 Ramm, S.A., 2007. Sexual selection and genital evolution in mammals: a phylogenetic analysis of baculum length. *The American Naturalist*, 169(3), pp. 360–69. Fitzpatrick, J.L., Almbro, M., Gonzalez-Voyer, A., Kolm, N. and Simmons, L.W., 2012. Male contest competition and the coevolution of weaponry and testes in pinnipeds. *Evolution*, 66, pp. 3595–604.
16 Foo, Y.Z., Simmons, L.W., Peters, M. and Rhodes, G., 2018. Perceived physical strength in men is attractive to women but may come at a cost to ejaculate quality. *Animal Behaviour*, 142, pp. 191–7.
17 Gwynne, D.T., 1993. Food quality controls sexual selection in Mormon crickets by altering male mating investment. *Ecology*, 74(5), pp. 1406–13.

18 Okuda, N., 1999. Sex roles are not always reversed when the potential reproductive rate is higher in females. *American Naturalist*, 153, pp. 540–48.
19 Geertjes, G.J. and Videler, J.J., 2002. A quantitative assessment of the reproductive system of the Mediterranean cave-dwelling triplefin blenny *Tripterygion melanurus*. *Marine Ecology*, 23, pp. 327–40.
20 Tobias, J.A. and Seddon, N., 2009. Signal jamming mediates sexual conflict in a duetting bird. *Current Biology*, 19, pp. 577–82.
21 Ma, W., Miao, Z. and Novotny, M.V., 1998. Role of the adrenal gland and adrenal-mediated chemosignals in suppression of estrus in the house mouse: the Lee-Boot effect revisited. *Biology of Reproduction*, 59(6), pp. 1317–20.
22 Young, A.J. et al., 2006. Stress and the suppression of subordinate reproduction in cooperatively breeding meerkats. *Proceedings of the National Academy of Sciences USA*, 103, pp. 12005–10.
23 Di Fiore, A. and Fleischer, R.C., 2005. Social behavior, reproductive strategies, and population genetic structure of *Lagothrix poeppigii*. *International Journal of Primatology*, 26, pp. 1137–73.
24 Doran-Sheehy, D.M., Fernández, D. and Borries, C., 2009. The strategic use of sex in wild female western gorillas. *American Journal of Primatology*, 71, pp. 1011–20.
25 Pusey, A.E. and Schroepfer-Walker, K., 2013. Female competition in chimpanzees. *Philosophical Transactions of the Royal Society B: Biological Sciences*, 368(1631), p. 20130077.
26 Ebensperger, L.A., 1998. Strategies and counterstrategies to infanticide in mammals. *Biological Reviews*, 73, pp. 321–46.
27 Tamura, N., 1995. Postcopulatory mate guarding by vocalization in the Formosan squirrel. *Behavioral Ecology and Sociobiology*, 36, pp. 377–86.
28 Foellmer, M.W. and Fairbairn, D.J., 2003. Spontaneous male death during copulation in an orb-weaving spider. *Proceedings of the Royal Society of London. Series B: Biological Sciences*, 270(Suppl_2), pp. S183–S185. Kuntner, M., Agnarsson, I. and Li, D., 2015. The eunuch phenomenon: adaptive evolution of genital emasculation in sexually dimorphic spiders. *Biological Reviews*, 90, pp. 279–96.
29 Komdeur, J., Kraaijeveld-Smit, F., Kraaijeveld, K. and Edelaar, P., 1999. Explicit experimental evidence for the role of mate guarding in minimizing loss of paternity in the Seychelles warbler. *Proceedings of the Royal Society of London. Series B: Biological Sciences*, 266(1433), pp. 2075–81.

Notes

30 Eggert, A.-K. and Sakaluk, S.K., 1995. Female-coerced monogamy in burying beetles. *Behavioral Ecology and Sociobiology*, 37(3), pp. 147–53.
31 Hare, R.M. and Simmons, L.W., 2019. Sexual selection and its evolutionary consequences in female animals. *Biological Reviews*, 94(3), pp. 929–56.
32 Alexander, R.D., Hoogland, J.L., Howard, R.D., Noonan, K.M. and Sherman, P.W., 1979. Sexual dimorphisms and breeding systems in pinnipeds, ungulates, primates, and humans. In: Chagnon, N.A. and Irons, W. (eds), *Evolutionary Biology and Human Social Behavior* (Scituate, MA: Duxbury Press), pp. 402–35. Pheasant, S.T., 1983. Sex differences in strength – some observations on their variability. *Applied Ergonomics* 14:205–11.
33 Lovejoy, C.O., 2009. Reexamining human origins in light of *Ardipithecus ramidus*. *Science*, 326, pp. 74e1–748e. Marlowe, F.W. and Berbesque, J.C., 2012. The human operational sex ratio: effects of marriage, concealed ovulation, and menopause on mate competition. *Journal of Human Evolution*, 63, pp. 834–42. Puts, D.A., Bailey, D.H. and Reno, P.L., 2015. Contest competition in men. In: Buss, D.M. (ed.), *The Handbook of Evolutionary Psychology* (Hoboken, NJ: Wiley & Sons), p. 526. Plavcan, J.M., 2012. Sexual size dimorphism, canine dimorphism, and male–male competition in primates: where do humans fit in? *Human Nature*, 23, pp. 45–67.
34 Kordsmeyer, T.L. et al., 2018. The relative importance of intra- and intersexual selection on human male sexually dimorphic traits. *Evolution and Human Behavior*, 39(4), pp. 424–36. Scott, I.M. et al., 2014. Human preferences for sexually dimorphic faces may be evolutionarily novel. *Proceedings of the National Academy of Sciences USA*, 111, pp. 14388–93. Dixson, B.J. and Vasey, P.L., 2012. Beards augment perceptions of men's age, social status, and aggressiveness, but not attractiveness. *Behavioral Ecology*, 23, pp. 481–90. Puts, D.A. et al., 2007. Men's voices as dominance signals: vocal fundamental and formant frequencies influence dominance attributions among men. *Evolution and Human Behavior*, 28, pp. 340–44.
35 Dixon, A.F., 2009. *Sexual Selection and the Origins of Human Mating Systems* (Oxford: Oxford University Press).
36 Von Rueden, C.R. and Jaeggi, A.V., 2016. Men's status and reproductive success in 33 nonindustrial societies: Effects of subsistence, marriage system, and reproductive strategy. *Proceedings of the National Academy of Sciences*, 113(39), pp. 10824–9.

37 Hennighausen, C., Hudders, L., Lange, B.P. and Fink, H., 2016. What if the rival drives a Porsche? Luxury car spending as a costly signal in male intrasexual competition. *Evolutionary Psychology*, 14(4), p. 1474704916678217.

38 Plavcan, J.M., 2012. Sexual size dimorphism, canine dimorphism, and male–male competition in primates: where do humans fit in? *Human Nature*, 23, pp. 45–67.

39 Deaner, R.O. and Smith, B.A., 2013. Sex differences in sports across 50 societies. *Cross-Cultural Research*, 47, pp. 268–309.

40 Brown, G.R., Laland, K.N. and Borgerhoff Mulder, M., 2009. Bateman's principles and human sex roles. *Trends in Ecology & Evolution*, 24, pp. 297–304.

41 Gat, A., 2015. Proving communal warfare among hunter-gatherers: the quasi-Rousseauan error. *Evolutionary Anthropology*, 24, pp. 111–26.

42 Betzig, L.L., 1986. *Despotism and Differential Reproduction: A Darwinian View of History* (Hawthorne, NY: Aldine de Gruyter).

43 Daly, M. and Wilson, M., 1988. *Homicide* (New Brunswick, NJ: Transaction Publishers). Daly, M. and Wilson, M., 1990. Killing the competition: female/female and male/male homicide. *Human Nature*, 1, pp. 81–107. Archer, J., 2009. Does sexual selection explain human sex differences in aggression? *Behavioral and Brain Sciences*, 32(3–4), pp. 249–66.

44 Chagnon, N.A., 1988. Life histories, blood revenge, and warfare in a tribal population. *Science*, 239(4843), pp. 985–92. Walker, R.S. and Bailey, D.H., 2013. Body counts in lowland South American violence. *Evolution and Human Behavior*, 34(1), pp. 29–34.

45 Pinker, S., 2012. *The Better Angels of Our Nature: Why Violence Has Declined* (New York: Penguin Books).

46 Keys, E. and Bhogal, M.S., 2016. Mean girls: provocative clothing leads to intra-sexual competition between females. *Current Psychology*, 37(3), pp. 543–51.

47 Milhausen, R.R. and Herold, E.S., 1999. Does the sexual double standard still exist? Perceptions of university women. *Journal of Sex Research*, 36(4), pp. 361–8.

48 Campbell, A., 2004. Female competition: causes, constraints, content and contexts. *Journal of Sex Research*, 41(1), pp. 16–26.

49 Benenson, J.F. and Abadzi, H., 2020. Contest versus scramble competition: sex differences in the quest for status. *Current Opinion in Psychology*, 33, pp. 62–8.

50 Campbell, A., 2004. Female competition: causes, constraints, content

and contexts. *Journal of Sex Research*, 41(1), pp. 16–26. Vaillancourt, T., 2013. Do human females use indirect aggression as an intrasexual competition strategy? *Philosophical Transactions of the Royal Society B: Biological Sciences*, 368(1631), p. 20130080.
51 Leenaars, L.S., Dane, A.V. and Marini, Z.A., 2008. Evolutionary perspective on indirect victimization in adolescence: the role of attractiveness, dating and sexual behavior. *Aggressive Behavior*, 34(4), pp. 404–15.
52 Silk, J.B., Alberts, S.C. and Altmann, J., 2003. Social bonds of female baboons enhance infant survival. *Science*, 302(5648), pp. 1231–4. Fratellone, G.P., Li, J.H., Sheeran, L.K., Wagner, R.S., Wang, X. and Sun, L., 2019. Social connectivity among female Tibetan macaques (*Macaca thibetana*) increases the speed of collective movements. *Primates*, 60, pp. 183–9.

Chapter 7: A Taste for the Beautiful – *Mate Choice*

1 Jin, S., 2020. Plastic's past: the complex historical and cultural influences underlying South Korea's plastic surgery phenomenon. *Synergy: The Journal of Contemporary Asian Studies*, 7. Available at: utsynergyjournal.org/2020/10/07/plastics-past-the-complex-historical-and-cultural-influences-underlying-south-koreas-plastic-surgery-phenomenon/ [accessed 27 August 2025].
2 *The Echo*, 2024. The case of the Korean cosmetic surgery industry. Available at: www.theechonews.com/article/2024/03/students-reflect-on-experiences [accessed 27 April 2025].
3 Petrie, M. and Halliday, T., 1994. Experimental and natural changes in the peacock's (*Pavo cristatus*) train can affect mating success. *Behavioral Ecology and Sociobiology*, 35, pp. 213–17.
4 Takahashi, M., Arita, H., Hiraiwa-Hasegawa, M. and Hasegawa, T., 2008. Peahens do not prefer peacocks with more elaborate trains. *Animal Behaviour*, 75(4), pp. 1209–19.
5 Dakin, R. and Montgomerie, R., 2011. Peahens prefer peacocks displaying more eyespots, but rarely. *Animal Behaviour*, 82(1), pp. 21–8.
6 Yorzinski, J.L., Patricelli, G.L., Babcock, J.S., Pearson, J.M. and Platt, M.L., 2013. Through their eyes: selective attention in peahens during courtship. *Journal of Experimental Biology*, 216(16), pp. 3035–46.
7 Wiley, R.H., 2015. *Noise Matters: The Evolution of Communication* (Cambridge, MA: Harvard University Press).
8 The stories, facts and quotes about Mike Ryan are based on my own

experience, interviews with him, and his autobiographic reflection: An improbable path. In: L. Drickamer and D. Dewsbury (eds), *Leaders in Animal Behavior, The Second Generation* (Cambridge: Cambridge University Press), pp. 465–96.

9 Petrie, M., 1994. Improved growth and survival of offspring of peacocks with more elaborate trains. *Nature*, 371(6498), pp. 598–9.

10 Kokko, H., Brooks, R., Jennions, M. and Morley, J., 2003. The evolution of mate choice and mating biases. *Proceedings of the Royal Society B: Biological Sciences*, 270(1515), pp. 653–64. Lewis, S. and South, A., 2012. The evolution of animal nuptial gifts. In: *Advances in the Study of Behavior* (Vol. 44, pp. 53–97) (New York: Academic Press).

11 Östlund-Nilsson, S., 2000. Female choice and paternal care in the fifteen-spined stickleback, *Spinachia spinachia*. PhD thesis, Acta Universitatis Upsaliensis.

12 Achorn, A.M. and Rosenthal, G.G., 2020. It's not about him: mismeasuring 'good genes' in sexual selection. *Trends in Ecology & Evolution*, 35(3), pp. 206–19.

13 Pike, T.W., Blount, J.D., Bjerkeng, B., Lindström, J. and Metcalfe, N.B., 2007. Carotenoids, oxidative stress and female mating preference for longer-lived males. *Proceedings of the Royal Society B: Biological Sciences*, 274(1618), pp. 1591–6.

14 Simons, M.J.P., Maia, R., Leenknegt, B. and Verhulst, S., 2014. Carotenoid-dependent signals and the evolution of plasma carotenoid levels in birds. *The American Naturalist*, 184(6), pp. 741–51. Weaver, R.J., Santos, E.S.A., Tucker, A.M., Wilson, A.E. and Hill, G.E., 2018. Carotenoid metabolism strengthens the link between feather coloration and individual quality. *Nature Communications*, 9(1), p. 345.

15 Dawkins, M. and Guilford, T., 1996. Sensory bias and the adaptiveness of female choice. *The American Naturalist*, 148(5), pp. 937–42. Andersson, M. and Simmons, L.W., 2006. Sexual selection and mate choice. *Trends in Ecology and Evolution*, 21(6), pp. 296–302.

16 Drăgănoiu, T.I., Nagle, L. and Kreutzer, M., 2002. Directional female preference for an exaggerated male trait in canary (*Serinus canaria*) song. *Proceedings of the Royal Society of London B: Biological Sciences*, 269(1509), pp. 2525–31. Pfaff, J.A., Zanette, L., MacDougall-Shackleton, S.A. and MacDougall-Shackleton, E.A., 2007. Song repertoire size varies with HVC volume and is indicative of male quality in song sparrows (*Melospiza melodia*). *Proceedings of the Royal Society B: Biological Sciences*, 274(1621), pp. 2035–40. Fuxjager, Matthew J. and

Barney A. Schlinger. Perspectives on the evolution of animal dancing: a case study of manakins. *Current Opinion in Behavioral Sciences* 6 (2015), pp. 7–12.

17   Keagy, J., Savard, J.F. and Borgia, G., 2012. Cognitive ability and the evolution of multiple behavioral display traits. *Behavioral Ecology*, 23(2), pp. 448–56. Chen, J., Zou, Y., Sun, Y.H. and Ten Cate, C., 2019. Problem-solving males become more attractive to female budgerigars. *Science*, 363(6423), pp. 166–7.

18   Yamazaki, K. and Beauchamp, G.K., 2007. Genetic basis for MHC-dependent mate choice. *Advances in Genetics*, 59, pp. 129–45. Bonneaud, C., Chastel, O., Federici, P., Westerdahl, H. and Sorci, G., 2006. Complex MHC-based mate choice in a wild passerine. *Proceedings of the Royal Society B: Biological Sciences*, 273(1590), pp. 1111–16. Puurtinen, M., Ketola, T. and Kotiaho, J., 2005. Genetic compatibility and sexual selection. *Trends in Ecology & Evolution*, 20(4), pp. 157–8.

19   Wedekind, C., Seebeck, T., Bettens, F. and Paepke, A.J., 1995. MHC-dependent mate preferences in humans. *Proceedings of the Royal Society of London. Series B: Biological Sciences*, 260(1359), pp. 245–9.

20   Gouda-Vossos, A., Nakagawa, S., Dixson, B.J.W. and Brooks, R.C., 2018. Mate choice copying in humans: a systematic review and meta-analysis. *Adaptive Human Behavior and Physiology*, 4(4), pp. 364–86.

21   Dean, R., Nakagawa, S. and Pizzari, T., 2011. The risk and intensity of sperm ejection in female birds. *The American Naturalist*, 178(3), pp. 343–54.

22   Kekäläinen, J. and Evans, J.P., 2018. Gamete-mediated mate choice: towards a more inclusive view of sexual selection. *Proceedings of the Royal Society B: Biological Sciences*, 285, p. 20180836.

23   Orr, T.J. and Zuk, M., 2014. Reproductive delays in mammals: an unexplored avenue for post-copulatory sexual selection. *Biological Reviews*, 89, pp. 889–912. Firman, R.C., Gasparini, C., Manier, M.K. and Pizzari, T., 2017. Postmating female control: 20 years of cryptic female choice. *Trends in Ecology & Evolution*, 32, pp. 368–82.

24   Byrne, P.G. and Rice, W.R., 2006. Evidence for adaptive male mate choice in the fruit fly *Drosophila melanogaster*. *Proceedings of the Royal Society of London B: Biological Sciences*, 273, pp. 917–22.

25   Richardson, J. and Zuk, M., 2024. Meta-analytical evidence that males prefer virgin females. *Ecology Letters*, 27(1), p. e14341.

26   Vincent, A., Ahnesjö, I., Berglund, A. and Rosenqvist, G., 1992. Pipefish and seahorses: are they all sex role reversed? *Trends in Ecology and Evolution*, 7(7), pp. 237–41.

27 Funk, D.H. and Tallamy, D.W., 2000. Courtship role reversal and deceptive signals in the long-tailed dance fly, *Rhamphomyia longicauda*. *Animal Behaviour*, 59(2), pp. 411–21.
28 Amundsen, T. and Forsgren, E., 2001. Male mate choice selects for female coloration in a fish. *Proceedings of the National Academy of Sciences*, 98, pp. 13155–60. Svensson, P.A., Forsgren, E., Amundsen, T. and Sköld, H.N., 2005. Chromatic interaction between egg pigmentation and skin chromatophores in the nuptial coloration of female two-spotted gobies. *Journal of Experimental Biology*, 208, pp. 4391–7.
29 Cunha, M., Berglund, A., Mendes, S. and Monteiro, N., 2018. The 'Woman in Red' effect: pipefish males curb pregnancies at the sight of an attractive female. *Proceedings of the Royal Society B*, 285(1885), p. 20181335.
30 Kurtovic, A., Widmer, A. and Dickson, B.J., 2007. A single class of olfactory neurons mediates behavioural responses to a *Drosophila* sex pheromone. *Nature*, 446(7135), pp. 542–6.
31 Sun, L., Wilczynski, W., Rand, A.S. and Ryan, M.J., 2000. Trade-off in short- and long-distance communication in tungara (*Physalaemus pustulosus*) and cricket (*Acris crepitans*) frogs. *Behavioral Ecology*, 11(1), pp. 102–9.
32 Renoult, J.P., Schaefer, H.M., Sallé, B. and Charpentier, M.J.E., 2011. The evolution of the multicoloured face of mandrills: insights from the perceptual space of colour vision. *PLoS ONE*, 6(12), e29117.
33 Boul, K.E., Funk, W.C., Darst, C.R., Cannatella, D.C. and Ryan, M.J., 2007. Sexual selection drives speciation in an Amazonian frog. *Proceedings of the Royal Society B: Biological Sciences*, 274(1608), pp. 399–406.
34 Andersson, M. and Simmons, L.W., 2006. Sexual selection and mate choice. *Trends in Ecology and Evolution*, 21(6), pp. 296–302.
35 Rosenthal, G.G., 2017. *Mate Choice: The Evolution of Sexual Decision Making from Microbes to Humans* (Princeton, NJ: Princeton University Press).
36 Bakker, T.C., 1993. Positive genetic correlation between female preference and preferred male ornament in sticklebacks. *Nature*, 363(6426), pp. 255–7. Cotton, S., Rogers, D.W., Small, J., Pomiankowski, A. and Fowler, K., 2006. Variation in preference for a male ornament is positively associated with female eyespan in the stalk-eyed fly *Diasemopsis meigenii*. *Proceedings of the Royal Society B: Biological Sciences*, 273(1591), pp. 1287–92.

Notes

37 Zahavi, A., 1975. Mate selection – a selection for a handicap. *Journal of Theoretical Biology*, 53(1), pp. 205–14.
38 Hill, G.E., 1991. Plumage coloration is a sexually selected indicator of male quality. *Nature*, 350(6316), pp. 337–9.
39 Petrie, M., 1994. Improved growth and survival of offspring of peacocks with more elaborate trains. *Nature*, 371(6498), pp. 598–9.
40 Portugal, R.D. and Svaiter, B.F., 2011. Weber–Fechner law and the optimality of the logarithmic scale. *Minds and Machines*, 21, pp. 73–81.
41 Akre, K.L., Farris, H.E., Lea, A.M., Page, R.A. and Ryan, M.J., 2011. Signal perception in frogs and bats and the evolution of mating signals. *Science*, 333(6043), pp. 751–2.
42 Buss, D.M., 2016. *Evolutionary Psychology: The New Science of Mind* (New York: Routledge).
43 Frederick, D.A. and Haselton, M.G., 2007. Why is muscularity sexy? Tests of the fitness indicator hypothesis. *Personality and Social Psychology Bulletin*, 33(8), pp. 1167–83. Puts, D.A., Jones, B.C. and DeBruine, L.M., 2012. Sexual selection on human faces and voices. *The Journal of Sex Research*, 49(2–3), pp. 227–43. Gildersleeve, K., Haselton, M.G. and Fales, M.R., 2014. Do women's mate preferences change across the ovulatory cycle? A meta-analytic review. *Psychological Bulletin*, 140, pp. 1205–59. Collins, S.A., 2000. Men's voices and women's choices. *Animal Behaviour*, 60(6), pp. 773–80. B.A. Scelza, Choosy but not chaste: multiple mating in human females, 2013. In Buss, D.M., 2016. *Evolutionary Psychology: The New Science of Mind* (New York: Routledge).
44 Rantala, M.J., Moore, F.R., Skrinda, I., Krama, T., Kivleniece, I., Kecko, S. and Krams, I., 2012. Evidence for the stress-linked immunocompetence handicap hypothesis in humans. *Nature Communications*, 3(1), p. 694. DeBruine, L.M., Little, A.C. and Jones, B.C., 2012. Extending parasite-stress theory to variation in human mate preferences. *Behavioral and Brain Sciences*, 35(2), pp. 86–7. DeBruine, L.M., Jones, B.C., Crawford, J.R., Welling, L.L.M. and Little, A.C., 2010. The health of a nation predicts their mate preferences: cross-cultural variation in women's preferences for masculinized male faces. *Proceedings of the Royal Society B: Biological Sciences*, 277(1692), pp. 2405–10. Jones, B.C., Feinberg, D.R., Watkins, C.D., Fincher, C.L., Little, A.C. and DeBruine, L.M., 2012. Pathogen disgust predicts women's preferences for masculinity in men's voices, faces, and bodies. *Behavioral Ecology*, 24(2), pp. 373–9.

45 Buss, D.M. and Schmitt, D.P., 2019. Mate preferences and their behavioral manifestations. *Annual Review of Psychology*, 70, pp. 77–110.
46 Marlowe, F.W., 2004. Mate preferences among Hadza hunter-gatherers. *Human Nature*, 15(4), pp. 365–76. Wang, G., Cao, M., Sauciuvenaite, J., Bissland, R., Hacker M. et al., 2018. Different impacts of resources on opposite sex ratings of physical attractiveness by males and females. *Evolution and Human Behaviour*, 39, pp. 220–25.
47 Betzig, L., 1989. Causes of conjugal dissolution: a cross-cultural study. *Current Anthropology*, 30(5), pp. 654–76.
48 Schmitt, D.P., 2005. Sociosexuality from Argentina to Zimbabwe: a 48-nation study of sex, culture, and strategies of human mating. *Behavioral and Brain Sciences*, 28(2), pp. 247–75. Lippa, R.A., 2009. Sex differences in sex drive, sociosexuality, and height across 53 nations: testing evolutionary and social structural theories. *Archives of Sexual Behaviour*, 38(5), pp. 631–51. Gray, P.B., 2013. Evolution and human sexuality. *American Journal of Physical Anthropology*, 152(S57), pp. 94–118. Buss, D.M. and Schmitt, D.P., 2019. Mate preferences and their behavioral manifestations. *Annual Review of Psychology*, 70, pp. 77–110.
49 Buss, D.M., 2016. *Evolutionary Psychology: The New Science of Mind* (New York: Routledge). Bode, A. and Kushnick, G., 2021. Proximate and ultimate perspectives on romantic love. *Frontiers in Psychology*, 12, p. 573123. Antfolk, J., 2017. Age limits: men's and women's youngest and oldest considered and actual sex partners. *Evolutionary Psychology*, 15(1), p. 1474704917690401. Antfolk, J., Salo, B., Alanko, K., Bergen, E., Corander, J., Sandnabba, N.K. and Santtila, P., 2015. Women's and men's sexual preferences and activities with respect to the partner's age: evidence for female choice. *Evolution and Human Behavior*, 36, pp. 73–9.
50 Buss, D.M., 1989. Sex differences in human mate preferences: evolutionary hypotheses tested in 37 cultures. *Behavioral and Brain Sciences*, 12, pp. 1–49. Prokosch, M.D., Coss, R.G., Scheib, J.E. and Blozis, S.A., 2009. Intelligence and mate choice: intelligent men are always appealing. *Evolution and Human Behavior*, 30(1), pp. 11–20.
51 Miller, G., 2011. *The Mating Mind: How Sexual Choice Shaped the Evolution of Human Nature* (New York: Anchor).
52 Dixson, B.J., Vasey, P.L., Sagata, K., Sibanda, N., Linklater, W.L. and Dixson, A.F., 2011. Men's preferences for women's breast morphology in New Zealand, Samoa, and Papua New Guinea. *Archives of Sexual Behavior*, 40, pp. 1271–9. Wheatley, J.R., Apicella, C.A., Burriss, R.P., Cardenas, R.A., Bailey, D.H., Welling, L.L.M. and Puts, D.A., 2014.

Women's faces and voices are cues to reproductive potential in industrial and forager societies. *Evolution and Human Behavior*, 35(4), pp. 264–71.
53 Ramsey, J.L., Langlois, J.H., Hoss, R.A., Rubenstein, A.J. and Griffin, A.M., 2004. Origins of a stereotype: categorization of facial attractiveness by 6-month-old infants. *Developmental Science*, 7(2), pp. 201–11.
54 Hahn, A.C. and Perrett, D.I., 2014. Neural and behavioral responses to attractiveness in adult and infant faces. *Neuroscience & Biobehavioral Reviews*, 46, pp. 591–603.
55 Ronay, R. and von Hippel, W. (2010). Power, risk, and the reckless male: testosterone and the effect of an audience on male risk-taking. *Evolution and Human Behavior*, 31(3), pp. 193–200.

Chapter 8: The Mating Game – *Sexual Conflict*
1 Klug, H., 2018. Why monogamy? A review of potential ultimate drivers. *Frontiers in Ecology and Evolution*, 6, p. 30.
2 Rose, C.M.S., Saucedo, H.A., Harper, C. and Jones, A.G., 2014. Genetic evidence for monogamy in the dwarf seahorse, *Hippocampus zosterae*. *The Journal of Heredity*, 105(6), pp. 828–927.
3 Fukuda, K., Manabe, H., Sakurai, M., Dewa, S., Shinomiya, A. and Sunobe, T., 2016. Monogamous mating system and sexuality in the gobiid fish, *Trimma marinae* (Actinopterygii: Gobiidae). *Journal of Ethology*, 35(1), pp. 51–6.
4 Griffith, S.C., Owens, I.P.F. and Thuman, K.A., 2002. Extra pair paternity in birds: a review of interspecific variation and adaptive function. *Molecular Ecology*, 11(11), pp. 2195–2212. Brouwer, L. and Griffith, S.C., 2019. Extra-pair paternity in birds. *Molecular Ecology*, 28(22), pp. 4864–82.
5 Ibid.
6 Taylor, M.L., Price, T.A.R. and Wedell, N., 2014. Polyandry in nature: a global analysis. *Trends in Ecology & Evolution*, 29(7), pp. 376–83.
7 Seeley, T.D. and Tarpy, D.R., 2007. Queen promiscuity lowers disease within honeybee colonies. *Proceedings of the Royal Society B: Biological Sciences*, 274(1606), pp. 67–72. Brouwer, L., Komdeur, J. and Richardson, D.S., 2007. Heterozygosity–fitness correlations in a bottlenecked island species: a case study on the Seychelles warbler. *Molecular Ecology*, 16(15), pp. 3134–44. Fossøy, F., Johnsen, A. and Lifjeld, J.T., 2008. Multiple genetic benefits of female promiscuity in a socially monogamous passerine. *Evolution*, 62(1), pp. 145–56.
8 Byrne, P.G. and Whiting, M.J., 2011. Effects of simultaneous polyandry

on offspring fitness in an African tree frog. *Behavioral Ecology*, 22(2), pp. 385–91.

9   Gerlach, N.M., McGlothlin, J.W., Parker, P.G. and Ketterson, E.D., 2012. Promiscuous mating produces offspring with higher lifetime fitness. *Proceedings of the Royal Society B: Biological Sciences*, 279(1730), pp. 860–66.

10 Rubenstein, D.R., 2007. Female extrapair mate choice in a cooperative breeder: trading sex for help and increasing offspring heterozygosity. *Proceedings of the Royal Society B: Biological Sciences*, 274(1620), pp. 1895–1903.

11 Tschirren, B., Postma, E., Rutstein, A.N. and Griffith, S.C., 2012. When mothers make sons sexy: maternal effects contribute to the increased sexual attractiveness of extra-pair offspring. *Proceedings of the Royal Society B: Biological Sciences*, 279(1731), pp. 1233–40.

12 Michl, G., Török, J., Griffith, S.C. and Sheldon, B.C., 2002. Experimental analysis of sperm competition mechanisms in a wild bird population. *Proceedings of the National Academy of Sciences*, 99(8), pp. 5466–70.

13 Paquet, M. and Smiseth, P.T., 2017. Females manipulate behavior of caring males via prenatal maternal effects. *Proceedings of the National Academy of Sciences*, 114(26), pp. 6800–6805.

14 Goyes Vallejos, J., Grafe, T.U. and Wells, K.D., 2018. Prolonged parental behaviour by males of *Limnonectes palavanensis* (Boulenger 1894), a frog with possible sex-role reversal. *Journal of Natural History*, 52(37–8), pp. 2473–85.

15 van Dijk, R.E., Szentirmai, I., Komdeur, J. and Székely, T., 2007. Sexual conflict over parental care in penduline tits *Remiz pendulinus*: the process of clutch desertion. *Ibis*, 149(3), pp. 530–34. Van Dijk, R.E., Székely, T., Komdeur, J., Pogany, Á., Fawcett, T.W. and Weissing, F.J., 2012. Individual variation and the resolution of conflict over parental care in penduline tits. *Proceedings of the Royal Society B: Biological Sciences*, 279(1735), pp. 1927–36.

16 Komdeur, J., Szentirmai, I., Székely, T., Bleeker, M. and Kingma, S.A., 2005. Body condition and clutch desertion in penduline tit *Remiz pendulinus*. *Behaviour*, 142(11–12), pp. 1465–78.

17 Moreno, J., Veiga, J.P., Cordero, P.J. and Mínguez, E., 1999. Effects of paternal care on reproductive success in the polygynous spotless starling *Sturnus unicolor*. *Behavioral Ecology and Sociobiology*, 47(1), pp. 47–53.

18 Donaldson, Z.R. and Young, L.J., 2008. Oxytocin, vasopressin, and the neurogenetics of sociality. *Science*, 322(5903), pp. 900–904.

Notes

19 Opie, C., Atkinson, Q.D., Dunbar, R.I.M. and Shultz, S., 2013. Male infanticide leads to social monogamy in primates. *Proceedings of the National Academy of Sciences*, 110(33), pp. 13328–32.
20 Swenson, J.E., Sandegren, F., Brunberg, S. and Wabakken, P., 1997. Infanticide caused by hunting of male bears. *Nature*, 386, pp. 450–51. Whitman, K., Starfield, A.M., Quadling, H.S. and Packer, C., 2004. Sustainable trophy hunting of African lions. *Nature*, 428, pp. 175–8.
21 Wolff, J.O. and Macdonald, D.W., 2004. Promiscuous females protect their offspring. *Trends in Ecology & Evolution*, 19(3), pp. 127–34.
22 Stockley, P., 2003. Female multiple mating behaviour, early reproductive failure and litter size variation in mammals. *Proceedings of the Royal Society of London. Series B: Biological Sciences*, 270, pp. 271–8.
23 Naim, D., Telfer, S., Sanderson, S., Kemp, S.J. and Watts, P.C., 2011. Prevalence of multiple mating by female common dormice, *Muscardinus avellanarius*. *Conservation Genetics*, 12, pp. 971–9.
24 Sun, L., 2023. *The Liars of Nature and the Nature of Liars* (Princeton, NJ: Princeton University Press).
25 Muller, M.N., Kahlenberg, S.M., Emery Thompson, M. and Wrangham, R.W., 2007. Male coercion and the costs of promiscuous mating for female chimpanzees. *Proceedings of the Royal Society B: Biological Sciences*, 274(1612), pp. 1009–14.
26 Fox, E.A., 2002. Female tactics to reduce sexual harassment in the Sumatran orangutan (*Pongo pygmaeus abelii*). *Behavioral Ecology and Sociobiology*, 52(2), pp. 93–101.
27 Muller, M.N., Kahlenberg, S.M., Emery Thompson, M. and Wrangham, R.W., 2007. Male coercion and the costs of promiscuous mating for female chimpanzees. *Proceedings of the Royal Society B: Biological Sciences*, 274(1612), pp. 1009–14.
28 Brennan, P.L.R., Clark, C.J. and Prum, R.O., 2010. Explosive eversion and functional morphology of the duck penis supports sexual conflict in waterfowl genitalia. *Proceedings of the Royal Society B: Biological Sciences*, 277(1686), pp. 1309–14. Brennan, P.L.R. and Prum, R.O., 2015. Mechanisms and evidence of genital coevolution: the roles of natural selection, mate choice, and sexual conflict. *Cold Spring Harbor Perspectives in Biology*, 7(7), p. a017749.
29 Brennan, P.L.R., Prum, R.O., McCracken, K.G., Sorenson, M.D., Wilson, R.E. and Birkhead, T.R., 2007. Coevolution of male and female genital morphology in waterfowl. *PLoS ONE*, 2, p. e418.
30 Chapman, T., Arnqvist, G., Bangham, J. and Rowe, L., 2003. Sexual conflict. *Trends in Ecology & Evolution*, 18(1), pp. 41–7.

31 Johnston, S.E., Gratten, J., Berenos, C., Pilkington, J.G., Clutton-Brock, T.H., Pemberton, J.M. and Slate, J., 2013. Life history trade-offs at a single locus maintain sexually selected genetic variation. *Nature*, 502(7469), pp. 93–5.
32 Sylvestre, F., Mérot, C., Normandeau, E. and Bernatchez, L., 2023. Searching for intralocus sexual conflicts in the three-spined stickleback (*Gasterosteus aculeatus*) genome. *Evolution*, 77(7), pp. 1667–81.
33 Haig, D. and Moore, T., 1991. Genomic imprinting in mammalian development: A parental tug-of-war. *Trends in Genetics*, 7(2), pp. 45–9. Butler, M.G., 2009. Genomic imprinting disorders in humans: a mini-review. *Journal of Assisted Reproduction and Genetics*, 26(9–10), pp. 477–86.
34 Ferguson-Smith, A.C. and Bourc'his, D., 2018. 2018 Gairdner Awards: The discovery and importance of genomic imprinting. *eLife*, 7, p. e42368.
35 Muralidhar, P., 2019. Mating preferences of selfish sex chromosomes. *Nature*, 570, pp. 376–9.
36 Brooks, R., 2000. Negative genetic correlation between male sexual attractiveness and survival. *Nature*, 406, pp. 67–70.
37 Reik, W. and Walter, J., 2001. Genomic imprinting: parental influence on the genome. *Nature Reviews Genetics*, 2(1), pp. 21–32.
38 Ferguson-Smith, A.C. and Bourc'his, D., 2018. 2018 Gairdner Awards: The discovery and importance of genomic imprinting. *eLife*, 7, p. e42368.
39 Creeth, H.D.J., McNamara, G.I., Tunster, S.J., Boque-Sastre, R., Allen, B., Sumption, L., Eddy, J.B., Isles, A.R. and John, R.M., 2018. Maternal care boosted by paternal imprinting in mammals. *PLoS Biology*, 16(7), p. e2006599.
40 Fillingim, R.B., King, C.D., Ribeiro-Dasilva, M.C., Rahim-Williams, B. and Riley, J.L., 2009. Sex, gender, and pain: a review of recent clinical and experimental findings. *The Journal of Pain*, 10(5), pp. 447–85.
41 Lassek, W.D. and Gaulin, S.J.C., 2009. Costs and benefits of fat-free muscle mass in men: Relationship to mating success, dietary requirements, and native immunity. *Evolution and Human Behavior*, 30(5), pp. 322–8.
42 National Cleveland Clinic, 2024. National Cleveland Clinic survey examines generational divide in men's health. [online] *Cleveland Clinic Newsroom*. Available at: https://newsroom.clevelandclinic.org/2024/09/04/national-cleveland-clinic-survey-examines-generational-divide-in-mens-health [accessed 12 May 2025].

43 Nunn, C.L., Lindenfors, P., Pursall, E.R. and Rolff, J., 2009. On sexual dimorphism in immune function. *Philosophical Transactions of the Royal Society B: Biological Sciences*, 364(1513), pp. 61–9.
44 Murdock, G.P., 1967. *Ethnographic Atlas* (Pittsburgh, PA: University of Pittsburgh Press). Alexander, R.D., Hoogland, J.L., Howard, R.D., Noonan, K.M. and Sherman, P.W., 1979. Sexual dimorphisms and breeding systems in pinnipeds, ungulates, primates, and humans. In: N.A. Chagnon and W. Irons (eds), *Evolutionary Biology and Human Social Behavior* (Scituate, MA: Duxbury Press), pp. 402–35.
45 Betzig, L., 1986. *Despotism and Differential Reproduction: A Darwinian View of History* (New York: Aldine).
46 Sun, L., 2013. *The Fairness Instinct: The Robin Hood Mentality and Our Biological Nature* (Amherst, NY: Prometheus Books).
47 Von Rueden, C., Gurven, M. and Kaplan, H., 2011. Why do men seek status? Fitness payoffs to dominance and prestige. *Proceedings of the Royal Society B: Biological Sciences*, 278, pp. 2223–32.
48 Scelza, B.A., 2012. Female choice and extrapair paternity in a traditional human population. *Biology Letters*, 7, pp. 889–91. Starkweather, K.E. and Hames, R., 2012. A survey of non-classical polyandry. *Human Nature*, 23, pp. 149–72. Beckerman, S., Lizarralde, R., Ballew, C., Schroeder, S., Fingelton, C., Garrison, A. and Smith, H., 1998. The Bari partible paternity project: preliminary results. *Current Anthropology*, 39, pp. 164–7.
49 Simmons, L.W., Firman, R.C., Rhodes, G. and Peters, M., 2004. Human sperm competition: testis size, sperm production and rates of extrapair copulations. *Animal Behaviour*, 68, pp. 297–302.
50 Anderson, K.G., 2006. How well does paternity confidence match actual paternity? Evidence from worldwide nonpaternity rates. *Current Anthropology*, 47, pp. 513–20. Scelza, B.A., Prall, S.P., Swinford, N., Goplan, S., Atkinson, E.G., McElreath, R., Sheehama, J. and Henn, B.M., 2020. High rate of extrapair paternity in a human population demonstrates diversity in human reproductive strategies. *Science Advances*, 6, eaay6195.
51 Faurie, C., Pontier, D. and Raymond, M., 2004. Student athletes claim to have more sexual partners than other students. *Evolution and Human Behavior*, 25(1), pp. 1–8.
52 Pillsworth, E.G. and Haselton, M.G., 2006. Male sexual attractiveness predicts differential ovulatory shifts in female extra-pair attraction and male mate retention. *Evolution and Human Behavior*, 27(4), pp. 247–58.

53 Ghanim, D., 2015. *The Virginity Trap in the Middle East* (New York: Springer). Mayeda, D.T. and Vijaykumar, R., 2016. A review of the literature on honor-based violence. *Sociology Compass*, 105(10), pp. 353–63.
54 Prokop, P. and Pazda, A.D., 2016. Women's red clothing can increase mate-guarding from their male partner. *Personality and Individual Differences*, 98, pp. 114–17.
55 Shackelford, T.K., Goetz, A.T., McKibbin, W.F. and Starratt, V.G., 2007. Absence makes the adaptations grow fonder: proportion of time apart from partner, male sexual psychology, and sperm competition in humans (*Homo sapiens*). *Journal of Comparative Psychology*, 121, pp. 214–20. Leivers, S., Rhodes, G. and Simmons, L.W., 2014. Sperm competition in humans: mate guarding behavior negatively correlates with ejaculate quality. *PLoS ONE*, 9(9), p. e108099. Pham, M.N. and Shackelford, T.K., 2015. Sperm competition and the evolution of human sexuality. In: T.K. Shackelford and R.D. Hansen (eds), *The Evolution of Sexuality* (Cham: Springer), pp. 257–75.
56 Goetz, A.T. and Shackelford, T.K., 2009. Sexual coercion in intimate relationships: a comparative analysis of the effects of women's infidelity and men's dominance and control. *Archives of Sexual Behavior*, 38, pp. 226–34.
57 Thornhill, R. and Palmer, C.T., 2000. *A Natural History of Rape* (Cambridge, MA: MIT Press). Gottschall, J., (2004). Explaining wartime rape. *Journal of Sex Research*, 41, 129–36.
58 Daly, M. and Wilson, M., 1988. *Homicide* (New York: Aldine de Gruyter).

Chapter 9: War and Peace in the Battle of the Sexes – *Sex and Evolutionary Destiny*

1 Fisher, R.A., 1930. *The Genetical Theory of Natural Selection*, Chapter 6: Sexual reproduction and sexual selection § Natural selection and the sex-ratio (Oxford: Clarendon Press), p. 141.
2 Basolo, A.L., 1994. The dynamics of Fisherian sex-ratio evolution: theoretical and experimental investigations. *The American Naturalist*, 144(3), pp. 473–90.
3 Caro, S.M., Griffin, A.S., Hinde, C.A. and West, S.A., 2016. Unpredictable environments lead to the evolution of parental neglect in birds. *Nature Communications*, 7(1), pp. 1–10.
4 Trivers, R.L. and Willard, D.E., 1973. Natural selection of parental ability to vary the sex ratio of offspring. *Science*, 179, pp. 90–92.
5 Cox, R.M. and Calsbeek, R., 2010. Cryptic sex-ratio bias provides

indirect genetic benefits despite sexual conflict. *Science*, 328(5974), pp. 92–4.
6   Pike, T.W. and Petrie, M., 2005. Maternal body condition and plasma hormones affect offspring sex ratio in peafowl. *Animal Behaviour*, 70(4), pp. 745–51.
7   Austad, S.N. and Sunquist, M.E., 1986. Sex-ratio manipulation in the common opossum. *Nature*, 324(6092), pp. 58–60.
8   Cameron, E.Z., 2004. Facultative adjustment of mammalian sex ratios in support of the Trivers–Willard hypothesis: evidence for a mechanism. *Proceedings of the Royal Society B: Biological Sciences*, 271(1549), pp. 1723–8.
9   Thogerson, C.M., Brady, C.M., Howard, R.D., Mason, G.J., Pajor, E.A., Vicino, G.A. and Garner, J.P., 2013. Winning the genetic lottery: biasing birth sex ratio results in more grandchildren. *PLoS ONE*, 8(7), e67867.
10 Komdeur, J., 1996. Facultative sex ratio bias in the offspring of Seychelles warblers. *Proceedings of the Royal Society B: Biological Sciences*, 263(1370), pp. 661–6.
11 Liker, A., Freckleton, R.P. and Székely, T., 2013. The evolution of sex roles in birds is related to adult sex ratio. *Nature Communications*, 4, p. 1587.
12 Cameron, E.Z., Edwards, A.M. and Parsley, L.M., 2017. Developmental sexual dimorphism and the evolution of mechanisms for adjustment of sex ratios in mammals. *Annals of the New York Academy of Sciences*, 1389(1), pp. 147–63.
13 Yang, X., Schadt, E.E., Wang, S. et al., 2006. Tissue-specific expression and regulation of sexually dimorphic genes in mice. *Genome Research*, 16(8), 995–1004.
14 Mitchell, N.J. and Janzen, F.J., 2010. Temperature dependent sex determination and contemporary climate change. *Sexual Development*, 4, pp. 129–40.
15 Montchamp-Moreau, C., 2006. Sex-ratio meiotic drive in *Drosophila simulans*: cellular mechanism, candidate genes and evolution. *Biochemical Society Transactions*, 34, pp. 562–5. Cocquet, J., Ellis, P.J., Mahadevaiah, S.K., Affara, N.A., Vaiman, D. et al., 2012. A genetic basis for a postmeiotic X versus Y chromosome intragenomic conflict in the mouse. *PLoS Genetics*, 8, e1002900.
16 Saumitou-Laprade, P., Cuguen, J. and Vernet, P., 1994. Cytoplasmic male sterility in plants: molecular evidence and the nucleocytoplasmic conflict. *Trends in Ecology & Evolution*, 9, pp. 431–5.
17 Kobayashi, S., Isotani, A., Mise, N. et al., 2006. Comparison of gene

expression in male and female mouse blastocysts revealed imprinting of the X-linked gene, *Rhox5/Pem*, at preimplantation stages. *Current Biology*, 16, pp. 166–72.

18 Rutkowska, J. and Badyaev, A.V., 2008. Meiotic drive and sex determination: molecular and cytological mechanisms of sex ratio adjustment in birds. *Philosophical Transactions of the Royal Society B: Biological Sciences*, 363, pp. 1675–86.

19 Goerlich-Jansson, V.C., Muller, M.S. and Groothuis, T.G.G., 2013. Manipulation of primary sex ratio in birds: lessons from the homing pigeon (*Columba livia domestica*). *Integrative and Comparative Biology*, 53, pp. 902–12.

20 Limbourg, T., Mateman, A.C. and Lessells, C.M., 2013. Opposite differential allocation by males and females of the same species. *Biology Letters*, 9, p. 20120835.

21 Grant, V.J., Irwin, R.J., Standley, N.T. et al., 2008. Sex of bovine embryos may be related to mothers' preovulatory follicular testosterone. *Biology of Reproduction*, 78, pp. 812–15.

22 Helle, S., Laaksonen, T., Adamsson, A. et al., 2008. Female field voles with high testosterone and glucose levels produce male-biased litters. *Animal Behaviour*, 75, pp. 1031–9. Shargal, D., Shore, L., Roteri, N. et al., 2008. Fecal testosterone is elevated in high-ranking female ibexes (*Capra nubiana*) and associated with increased aggression and a preponderance of male offspring. *Theriogenology*, 69, pp. 673–80. Grant, V.J. and Irwin, R.J., 2010. Can mammalian mothers influence the sex of their offspring peri-conceptually? *Reproduction*, 140, pp. 425–33.

23 Sheldon, B.C. and West, S.A., 2004. Maternal dominance, maternal condition, and offspring sex ratio in ungulate mammals. *The American Naturalist*, 163, pp. 40–54. Alvarez, L., Arvizu, R.R., Luna, J.A. and Zarco, L.A., 2010. Social ranking and plasma progesterone levels in goats. *Small Ruminant Research*, 90, pp. 161–4.

24 Firman, R.C., Tedeschi, J.N. and Garcia-Gonzalez, F., 2020. Sperm sex ratio adjustment in a mammal: perceived male competition leads to elevated proportions of female-producing sperm. *Biology Letters*, 16(6), p. 20190929.

25 James, W.H. and Grech, V., 2017. A review of the established and suspected causes of variations in human sex ratio at birth. *Early Human Development*, 109, pp. 50–56.

26 Peippo, J. and Bredbacka, P., 1995. Sex-related growth rate differences in mouse pre-implantation embryos in vivo and in vitro. *Molecular*

Reproduction and Development, 40, pp. 56–61. Kimura, K., Spate, L.D., Green, M.P. and Roberts, R.M., 2005. Effects of d-glucose concentration, d-fructose, and inhibitors of enzymes of the pentose phosphate pathway on the development and sex ratio of bovine blastocysts. *Molecular Reproduction and Development*, 72, pp. 201–7.

27 Helle, S., Laaksonen, T., Adamsson, A. et al., 2008. Female field voles with high testosterone and glucose levels produce male-biased litters. *Animal Behaviour*, 75, pp. 1031–9.

28 Cornwallis, C.K. and O'Connor, E.A., 2009. Sperm: seminal fluid interactions and the adjustment of sperm quality in relation to female attractiveness. *Proceedings of the Royal Society B: Biological Sciences*, 276, pp. 3467–75. Perry, J.C., Sirot, L. and Wigby, S., 2013. The seminal symphony: how to compose an ejaculate. *Trends in Ecology & Evolution*, 28, pp. 414–22. Firman, R.C., Tedeschi, J.N. and Garcia-Gonzalez, F., 2020. Sperm sex ratio adjustment in a mammal: perceived male competition leads to elevated proportions of female-producing sperm. *Biology Letters*, 16(6), p. 20190929.

29 Malo, A.F., Martinez-Pastor, F., Garcia-Gonzalez, F., Garde, J., Ballou, J.D. and Lacy, R.C., 2017. A father effect explains sex-ratio bias. *Proceedings of the Royal Society B: Biological Sciences*, 284(1861), p. 20171159.

30 Douhard, M. and Geffroy, B., 2021. Males can adjust offspring sex ratio in an adaptive fashion through different mechanisms. *BioEssays*, 43(5), p. 2000264.

31 Weatherhead, P.J. and Robertson, R.J., 1979. Offspring quality and the polygyny threshold: 'the sexy son hypothesis'. *The American Naturalist*, 113(2), pp. 201–8.

32 Gwinner, H. and Schwabl, H., 2005. Evidence for sexy sons in European starlings (*Sturnus vulgaris*). *Behavioral Ecology and Sociobiology*, 58(4), pp. 375–82.

33 Fabiani, A., Galimberti, F., Sanvito, S. and Hoelzel, A.R., 2004. Extreme polygyny among southern elephant seals on Sea Lion Island, Falkland Islands. *Behavioral Ecology*, 15, pp. 961–9.

34 Lovari, S., Pellizzi, B., Boesi, R. and Fusani, L., 2009. Mating dominance amongst male Himalayan tahr: blonds do better. *Behavioural Processes*, 81, pp. 20–25.

35 Rintamäki, P.T., Höglund, J., Alatalo, R.V. and Lundberg, A., 2001. Correlates of male mating success on black grouse (*Tetrao tetrix L.*) leks. *Annales Zoologici Fennici*, pp. 99–109.

36 Corl, A. and Ellegren, H., 2012. The genomic signature of sexual selection in the genetic diversity of the sex chromosomes and autosomes. *Evolution*, 66(7), pp. 2138–49.
37 Wilder, J.A., Mobasher, Z. and Hammer, M.F., 2004. Genetic evidence for unequal effective population sizes of human females and males. *Molecular Biology and Evolution*, 21, pp. 2047–57. Lohmueller, K.E., Degenhardt, J.D. and Keinon, A., 2010. Sex-averaged recombination rates on the X chromosome: a comment on Labuda et al. *American Journal of Human Genetics*, 86, pp. 978–81. Balaresque, P., Poulet, N., Cussat-Blanc, S., Gerard, P., Quintana-Murci, L., Heyer, E. and Jobling, M.A., 2015. Y-chromosome descent clusters and male differential reproductive success: young lineage expansions dominate Asian pastoral nomadic populations. *European Journal of Human Genetics*, 23, pp. 1413–22.
38 Byers, J.A. and Waits, L., 2006. Good genes sexual selection in nature. *Proceedings of the National Academy of Sciences*, 103(44), pp. 16343–4.
39 Cunningham, E.J.A. and Russell, A.F., 2000. Egg investment is influenced by male attractiveness in the mallard. *Nature*, 404(6773), pp. 74–7.
40 Gil, D., Leboucher, G., Lacroix, A., Cue, R. and Kreutzer, M., 2004. Female canaries produce eggs with greater amounts of testosterone when exposed to preferred male song. *Hormones and Behavior*, 45(1), pp. 64–70. Loyau, A., Saint Jalme, M., Mauget, R. and Sorci, G., 2007. Male sexual attractiveness affects the investment of maternal resources into the eggs in peafowl (*Pavo cristatus*). *Behavioral Ecology and Sociobiology*, 61(7), pp. 1043–52.
41 Pollet, T.V., Fawcett, T.W., Buunk, A.P. and Nettle, D., 2009. Sex-ratio biasing towards daughters among lower-ranking co-wives in Rwanda. *Biology Letters*, 5(6), pp. 765–8.
42 Hartung, J., 1982. Polygyny and inheritance of wealth. *Current Anthropology*, 23, pp. 1–12.
43 Cronk, L., 1993. Parental favoritism toward daughters. *American Scientist*, 81(3), pp. 272–9.
44 Catalano, R.A., 2003. Sex ratios in the two Germanies: a test of the economic stress hypothesis. *Human Reproduction*, 18(9), pp. 1972–5.
45 Kanazawa, S. and Vandermassen, G., 2005. Engineers have more sons, nurses have more daughters: an evolutionary psychological extension of Baron–Cohen's extreme male brain theory of autism. *Journal of Theoretical Biology*, 233(4), pp. 589–99.

46 Kanazawa, S., 2005. Big and tall parents have more sons: further generalizations of the Trivers–Willard hypothesis. *Journal of Theoretical Biology*, 235(4), pp. 583–90.
47 Cameron, E.Z. and Dalerum, F., 2009. A Trivers–Willard effect in contemporary humans: male-biased sex ratios among billionaires. *PLoS ONE*, 4(1), e4195.
48 Zerjal, T., Xue, Y., Bertorelle, G., Wells, R.S., Bao, W., Zhu, S., Qamar, R. et al., 2003. The genetic legacy of the Mongols. *The American Journal of Human Genetics*, 72(3), pp. 717–21.
49 Ibid.
50 Barash, D.P., 2002. Evolution, males, and violence. *The Chronicle of Higher Education*. Available at: www.chronicle.com/article/evolution-males-and-violence/ [accessed 22 May 2025].
51 Brown, G.R., Laland, K.N. and Borgerhoff Mulder, M., 2009. Bateman's principles and human sex roles. *Trends in Ecology & Evolution*, 24(6), pp. 297–304.
52 Hudson, V. and Den Boer, A.M., 2002. A surplus of men, a deficit of peace: security and sex ratios in Asia's largest states. *International Security*, 26(4), pp. 5–38.
53 Barber, N., 2003. The sex ratio and female marital opportunity as historical predictors of violent crime in England, Scotland, and the United States. *Cross-Cultural Research*, 37(4), pp. 373–92.
54 D'Alessio, S.J. and Stolzenberg, L., 2010. The sex ratio and male-on-female intimate partner violence. *Journal of Criminal Justice*, 38(4), pp. 555–61.
55 Edlund, L., Li, H., Yi, J. and Zhang, J., 2013. Sex ratios and crime: evidence from China. *Review of Economics and Statistics*, 95(5), pp. 1520–34.
56 McAuley, J.B., Servin, B., Burnett, H.A., Brekke, C., Peters, L., Hagen, I.J., Niskanen, A.K., Ringsby, T.H., Husby, A., Jensen, H. and Johnston, S.E., 2024. The genetic architecture of recombination rates is polygenic and differs between the sexes in wild house sparrows (*Passer domesticus*). *Molecular Biology and Evolution*, 41(9), p. msae179.
57 Lenormand, T., Engelstädter, J., Johnston, S.E., Wijnker, E. and Haag, C.R., 2016. Evolutionary mysteries in meiosis. *Philosophical Transactions of the Royal Society B: Biological Sciences*, 371(1706), p. 20160001. Sardell, J.M. and Kirkpatrick, M., 2020. Sex differences in the recombination landscape. *The American Naturalist*, 195(2), pp. 361–79. Johnston, S.E., 2024. Understanding the genetic basis of variation in meiotic recombination: past, present, and future. *Molecular Biology and Evolution*, 41(7), p. msae112.

IMAGE PERMISSIONS

0.1. Ciliated protist *T. thermophila*, image courtesy of Ed Orias
1.1. California condor close-up, photograph by Chuck Szmurlo via Wikimedia Commons (left) and California condor flying (right) with permission from Michael J. Parr, American Bird Conservancy
1.2. Meiosis overview by Rdbickel via Wikimedia Commons
1.3. Bacterial conjugation by Jonasz Patkowski via Wikimedia Commons
1.4. Bdelloid rotifers micrograph courtesy of Diego Fontaneto, *PLoS Biology* Journal
1.5. Phylogenetic tree of stick insects, adapted from K. S. Jaron et al., in *Convergent consequences of parthenogenesis on stick insect genomes*, *Science Advances*
2.1. Red Queen dynamics, diagram by author
2.2. Endangered black-footed ferrets, photograph by Kimberly Fraser via Creative Commons
2.3. Snowshoe hare–lynx dynamic, diagram by author
3.1. August Weismann's germ plasm theory, diagram by Ian Alexander via Wikimedia Commons
3.2. Three main types of life cycles, diagram by author
3.3. Mushroom life cycle, diagram modified from C. Lull et al., in *Antiinflammatory and immunomodulating properties of fungal metabolites*, *Mediators of Inflammation* via Wiley
3.4. A cluster of mushrooms growing on a mossy ground, photograph via Pixabay
3.5. Isogamy verses anisogamy, diagram by Tameeria and Helix84 via Wikimedia Commons
3.6. The split-gill mushroom, photo courtesy of Orien Sun
3.7. Male ruffs, image by Loveland et al., in *A single gene orchestrates androgen variation underlying male mating morphs in ruffs*, *Science*

Image Permissions

Journal, courtesy of American Association for the Advancement of Science

4.1. Sex chromosomes in mammals and birds, diagram by author

4.2. Platypus artwork by Nellie Pease/CABAH via Wikimedia Commons (right) and platypus chromosomes image from Grützner et al., *Genomic Biology* Journal courtesy of Springer Nature

4.3. Haplodiploid system in bees, diagram by author

4.4. Snow skink in Tasmania, photograph by dhfischer via iNaturalist

4.5. Swamp eel, photograph by Sakdinon via Research Gate

4.6. Barr bodies in female cells, image © Steven M. Carr

5.1. Jacanas photograph by Benjamin Keen (left) and Mdkshots (right) via Wikimedia Commons

5.2. Spotted hyena photograph by Bernard Dupont via Wikimedia Commons

5.3. Two-horned uterus from Ryan and Vandenburgh in *Intrauterine position effects, Neuroscience & Biobehavioral Reviews* via Science Direct

5.4. Human hand ratio photograph by author with help from Daya S. Brar

5.5. Female and male seahorse photograph by Elizabeth Haslam via Wikimedia Commons

6.1. Vertebrate weapons from Emlen in *The evolution of animal weapons* © Annual Reviews, Inc

6.2. Rattlesnakes fighting photograph by Dawn Endico via openverse

6.3. Sperm diversity from S. Lüpold and S. Pitnick in *Sperm form and function, Reproduction* Journal © Society for Reproduction and Fertility 2018

6.4. Horned beetle diagram from L. W. Simmons and D. J. Emlen, *Evolutionary trade-off between weapons and testes, PNAS* © 2006, National Academy of Sciences, USA

6.5. Garden spiders illustration by James Genry Emerton via Wikimedia Commons

7.1. Beauty and cosmetic surgery photograph courtesy of Sophie Jin, Memeburn

7.2. Blue peacock photograph via Pixabay

7.3. Guianan cock-of-the-rock photograph by Juniorgriotto via Wikimedia Commons

7.4. Female dance fly photograph from David Funk in *Courtship role reversal and deceptive signals in the long-tailed dance fly*, Rhamphomyia longicauda, *Animal Behaviour* via Science Direct © 2000 The

Association for the Study of Animal Behaviour and published by Elsevier Ltd

7.5. Mandrill photograph by Rolf Dietrich Brecher (left) and uakari photograph by Kevin O'Connel (right) via Wikimedia

7.6. Weber-Fechner example illustration by MrPomidor via Wikimedia Commons

7.7. Lincoln image by Anthony Berger, Mathew Brady Studio and Harding photograph by Harris & Ewing

8.1. Elk and sea elephant life histories diagram from T. H. Clutton-Brock in *Social evolution in mammals*, *Science* Journal via American Association for the Advancement of Science

8.2. Surinam toad illustration by Frederick P. Nodder (left) via Wikimedia Commons and smooth garden frogs photograph by Chien Lee (right), via Minden Pictures

8.3. Penduline tit artwork by Naumann via Wikimedia Commons

8.4. Paternal-maternal conflict in mouse embryos diagram from J. Ågren, & A. G. Clark in *Selfish genetic elements*, *PLoS Genetics* Journal

9.1 Sex ratios in platyfish from Basolo in *The dynamics of fisherian sex-ratio evolution*, *The American Naturalist*, courtesy of the American Society of Naturalists

9.2. A male tahr photograph courtesy of Department of National Parks and Wildlife Conservation, Nepal, via Wikimedia Commons

9.3. Table showing DNA compositions in Latin American populations from K. Adhikari et al., *Admixture in Latin America*, *Current Opinion in Genetics & Development* © 2016 Elsevier Ltd

INDEX

Page numbers in *italics* indicate illustrations

Aché people, Paraguay 278
Achorn, Angela 209
Adams, John 172–5, 229
Aesop 72
Agol, Isador 20
Agrawal, Aneil 32
algae 78, 81, 86
Allee, Warder Clyde 56
Allee effect 56–7
alligators 114, 146
allozyme electrophoresis 58
American Psychiatric Association 154
American Psychological Association 154
Ancient Greeks 128, 163, 273
androgens 129, 134, 144, 149–51, 154
  androgen insensitivity syndrome (AIS) 133–4
anisogamy 82, *83*, 86, 299
'Ant and the Grasshopper, The' (Aesop), 72
antbirds (*Hypocnemis peruviana*) 191
antelopes 155
  topi antelopes 213, 270
antibiotics 63
ants 107, *108*, 110–12, 126
apes 127
aphids 32, 38, 52–3, 86

Aristotle 128, 174
asexual reproduction 10–15, 21–31, 34–6, *30*, 86–7
  Muller's ratchet 16–17, 21–32, 34–6, 286–7, 302
autosomes 123, 277, 283, 299

baboons 155, 198
bacteria 24–8, 63
  conjugation 24–5, *25*, 28, 43, 299
  genomes 27
  transduction 24, 28, 303
  transformation 24, 28, 303
bacteriophages 24, 42, 303
Bailey, J. Michael 152
Bambach, Richard 95–6
Banks, Joseph 171
Barash, David 233, 278
Barash, Nanelle 233
Barr, Murray 133
Barr bodies, 133, *133*, 299
Barton, Sheila 250
Basolo, Alexandra 258–9
Bateman's rule 178, 183, 189, 197, 213, 225, 233, 254, 299
bats 73n, 270
'battle of the sexes' 70–73, 176, 284
bdelloid rotifers 25–6, *27*, 29, 43, 53

beauty 200–230, 273–4
beauty pageants 200–201, *201*, 228
Becks, Lutz 61–2
bees 107–10, *108*, 110, 126
beetles
   burying beetles 193–4, 238
   flour beetles 47
   scarab beetles 188, *188*
Belding's ground squirrel 127
Benenson, Joyce 197
Bernoulli, Jakob 38n
Bertram, Ewart George 133
Bethune, Norman 20
Betzig, Laura 225
biodiversity 96–7
biological sex 82–3, 125, 127–8, 141, 145, 292–3
birds 105–7, *105*, 121, 209–11, 260–61
   carotenoids and 209–10, *210*
   eggs 261, 262
   intelligence and 210–11
   monogamy and 236, 239
   parental care and 239–40, 261
   sex ratios and 266
   song 210
black-footed ferrets 51, *51*, 57
Blepharisma ciliate 84, 85
Bligh, William 171
body size 95
Bonellia viridis marine worms 117–18
bonnethead sharks 86
bonobos 155, 187
boubous 194
Boul, Kathy 217
Bounty, mutiny on the 171–5
Brachionus calyciflorus rotifer 62
Brazil 142
Brennan, Patricia 244
brown anole lizard 262
Brown, William 172–3

burying beetles
   Nicrophorus defodiens 193–4
   Nicrophorus vespilloides 238
Bush, Andrew 95–6
Buss, David 223–5

calico cats 249
California condors 9–11, *10*, 31, 51, 57, 86, 105, 176
canaries 210
cardinalfish 190
caribou 149–50
carotenoids 209–10
Carroll, Lewis 41
celibacy 26
cells 1–2, 74–5, 93–4
   meiosis 13, *14*, 22n, 34, 80, 266, 302
   gametes (sex cells) 71–5, 73, 75, 82–3, 87–8, 90, 94, 175, 280
   sperm cells 83–4, 86–7, 90, 93m 212–13
   see also diploids; haploids
Chakra, Maria 61–2
cheetahs 51
chickens 211–12
   Illinois prairie chickens 52
   roosters 176
chimpanzees 155, 187n, 191, 194, 243
China 128, 163, 274, 278, 279
Chlamydomonas green algae 86
Christian, Fletcher 171–3
chromosomes 14, 272
   autosomes 123, 277, 283, 299
   diploids 74, 78–81, *79*, 90–91, 205, 300
   haploids 74, 78–81, *79*, 90–91, 301
   sex chromosomes 104–8, *105*, *106*, 120–25, 129–35, 248–50, 302
   sex ratios and 265–6
   sexual orientation and 152–3

triploids 49, 58, 303
X-inactivation 249
ciliates 84, 85
  Blepharisma 84, 85
  Tetrahymena hyperangularis 84
Clark, Mertice 147–8
Cleopatra 273
climate change 117
cloning 11–12, 21, 34, 91–3, 286
  Dolly the sheep 34
  early life 69–70
  snails 57–8
clownfish 118–20, 141, 159, 190
Clutton-Brock, Tim 181
coalescence 269n, 282n
cognitive bias 228
collared flycatcher 236
Columbia University, New York 17
condition-dependent sex 32–3
condors 9–11, *10*, 31, 51, 57, 86, 105, 176
congenital adrenal hyperplasia (CAH) 129
conjugation 24–5, *25*, 28, 43, 299
cooperative bargaining 183, 291, 299
copulation rates 270
cortisol 129
'cost of meiosis' 13, 108–10, 112, 126
'cost of sex/cost of males' 13–14, 299–300
Craddock, Clark 50
crows 210
Cunha, Mario 213

Dakin, Roslyn 203
damselflies 186
dance flies 213, *214*
dandelions 47–8
Darwin, Charles 75, 178, 196, 202, 207, 214, 217, 218, 226n

Darwinian fitness 32, 99, 162, 247, 260
David, Alphonso 292–2
Dawkins, Richard 88–90, 95n, 125, 174, 179, 246, 255
Dawkins's vehicle, 90, 126, 162, 167, 187, 214, 257
Delph, Lynda 45–6, 48–9, 57, 64
demographic limitation 292
Descent of Man, The (Darwin) 178, 202, 214, 226n
Diaochan 274
Dictyostelium discoideum slime mould 84
diploids 74, 78–81, *79*, 90–91, 205, 300
disorders of sex development (DSD) 128–31
Dmrt1 gene 124
DNA 23–4, 272
  chloroplast DNA 23n
  damage/repair 22n, 35–6, 39, 43
  mitochondrial DNA 23n,
  nuclear mitochondrial DNA 24n
  recombination, 28, 32, 59
  transduction 24, 28, 303
  transformation 24, 28, 303
Dogon people, Mali 278
dolphins 155
ducks 244–5, 273
  mallards 273
  Muscovy 244–5
Dybdahl, Mark 58

East China Normal University, Shanghai 88
eggs 73–5, 86, 90, 176–7, 184–6, 189, 205
elephant seals 179, 188, 196, 234, *234*, 247, 251, 264, 270
elephants 155, 181

elks 203
emus 139
endangered species 51–2
environmental sex determination 113–17
enzymes 58
epigenetics 154
estradiol 148
eukaryotes 16, 26, 29
Eurasian penduline tits 239–40, *241*
European dunnocks 186
European moles 3, 150
eusociality 113
evolution 96–9, 273, 286–91
evolutionary biology 45, 71–7
Evolutionary Theory journal 64
extinction 51–2, 69
    extinction rates 41
    extinction vortex 56–7

'false mating' 34
Fechner, Gustav Theodor 220
Feldman, Marcus 64
Felsenstein, Joe 21
ferrets, black-footed 51, *51*, 57
fertility 149, 269, 288
fertilization 184–8, *185*, 211–12
    external 238–9
    internal 239
Finland 149
fish 121, 176
    see also names of individual fish
Fisher, Ronald A. 218–19, 222, 258–9, 261, 268–9
Fisher's runaway hypothesis 218–19, 258, 300
flatworms 32, 46
Flaubert, Gustave 231–2
Florida, USA 146
Florida panthers 52

Florida scrub jay 126–7
flour beetles 47
foetal development 147–51
Folger, Mayhew 173
Foo, Yong Zhi 189
Formosan squirrels 192
frequency-dependent selection 43, 259
Freud, Sigmund 223
frogs 121, 176, 206–8, 214–18, 221–2, 239
    cricket frogs 215
    Peters' dwarf frogs 217–18
    smooth guardian frogs, 239
    túngara frogs 215, 221–2
fruit flies (*Drosophila*) 17–18, 21, 47, 105, 213, 214
Fuentes, Agustín 165
fungi 78–81, 84–5, *80*, *82*
    common (*Agaricus bisporus*) 84
    fairy inkcap (*Coprinellus disseminatus*) 84
    honey mushrooms (*Armillaria ostoyae*) 81–2, *82*
    mycelia 79, 81
    split-gill (*Schizophyllum commune*) 84–5, *84*

Galdikas, Biruté 243
Galef, Bennett 147–8
gambling 37–9
gametes (sex cells) 71–5, 73, 75, 82–3, 87–8, 90, 94, 175, 280
    see also haploids
garden spiders 192, *193*
garter snakes 186, 192
Ge, Chutian 114
geckos 114
    Heteronotia binoei 47
geese 155

Index

gender 140–46, 151–67, 289–90, 300
  bending gender 4, 139, 143–4, 160–61, 290
  gender reversal, 4, 118–20, 141, 158–62, 213, 264, 287, 289–90, 300
  hormones and 146–51
  human gender roles 162–6
  role reversal, 137–45, 160–61, 300
  see also sexual orientation
genetics 70–74, 78, 91–4, 256–8
  genes 88–91, 93–5
  genetic collectivism 92
  genetic diversity 59, 63, 70, 91–2
  genetic individualism 92
  genetic recombination, 28, 32, 59
  genetic relatedness 107–12
  genetic rescue 52, 300
  horizontal gene transfer 24–5
  master genes 120–23, 302
  recombination 17, 32, 35, 65, 88–9, 92, 122, 174, 280–83, 287–8, 300
  sex and 93–5
  sex determination and 104–13, 120–31
  sex ratios and 265–6, 272, 276–8
  sexual conflict and 245–51, 247
  sexual orientation and 152–3
  t-complex genes 93
  see also chromosomes, DNA; genomes
Genghis Khan, 276, 282
genomes 22–4, 74, 94
  bacterial 27
  genomic imprinting 245, 300
genotypes 43, 57–8
Germany 19, 275
giraffes 76, 181
goby fish 119–20, 141, 213
  princess pygmy goby 235
golden-collared manakins 210

Goodall, Jane 206
gorillas 179, 191, 194, 196, 203
Gray, Asa 202
great tits 194
grouse 270
  black grouse 271
Grützner, Frank 107
Guianan cock-of-the-rock, *210*
Guillette Jr, Louis 146
guppies 183, 248–9, 268

Haafke, Julia 61–2
Hadany, Lilach 32
Haig, David 94
'halo effect' 228, 274
Hamer, Dean 167
handicap hypothesis, 219–20, 300
haploids 74, 78–81, *79*, 90–91, 301
  haplo-diploid sex determination system 110–132
  haploid gametes 205
Harding, Warren G. 229, *230*
'height premium' 228–9
hermaphrodites 31n, 86n, 301
Heteronotia binoei gecko 47
Himalayan tahr 270, *271*
Himba people, Namibia 252
Hippocrates 147
Hitler, Adolf 19
Hobbes, Thomas 174, 182, 183, 198
homosexuality 151–5
hormones 124, 129, 134, 144, 146–51, *147*
  androgens 129, 134, 144, 149–51, 154
  anti-Müllerian hormones 124
  cortisol 129
  estradiol 148
  oestrogen 146, 148
  sex ratios and 266–8

sexual conflict and 241
sexual orientation and 154–5
testosterone 133–5, 146, 148–51, 154–5, 266–8, 224
horses 245–6, 248
house finches 219
humans
  eggs 74, 87
  fingers 151, *151*
  foetal development, 148–9
  gender 151–2, 162–4
  homicide and 196–7, 278
  infanticide 254–5
  infidelity 253–4
  intersex individuals 128–31, *130*
  kin selection 113, 156–7
  male violence 195–6, 278–9
  mate choice 223–30, 252–3
  mate competition 194–8
  monogamy 252–3
  sex determination 121–5
  sex identity 131–5
  sex ratios 273–80
  sexual coercion 253–4
  sexual conflict 251–5
  sexual orientation 151–9
  sperm 74, 87
Hunt, Gene 95–6
hyenas 176
  spotted hyenas 143–5, *145*, 213
hypergamy 224n
hypospadias 128–9

Iberian lynx 57
iguanas 114
Illinois prairie chickens 52
immune system 59
inbreeding 50–53, 85, 112
inbreeding depression 51, 301
Incas 163

India 163, 279
infanticide 183, 191, 242, 254–5
insects 106–8, 121, 155
International Association of Athletics Federations (IAAF) 131–2
International Olympic Committee (IOC) 132–4
intersex 4, 83, 109, 117, 126–31, *130*, 134, 156, 289, 293, 301
Isle Royale National Park, Lake Superior 52
Isoetes (quillworts) 60
isogamy 83, *83*, 301
ivorybilled woodpecker 56

Jablonski, David 64
jacanas 137–9, 141, *139*, 160–61, 176, 190, 213, 264, 290
Jokela, Jukka 60

K-strategists 55, 301
Kant, Immanuel 183
Kennedy, John 292–3
Khelif, Imane 134–5
kin selection 13, 83, 109–13, 126–7, 156–7, 288–9, 301
Kipsigis people, Kenya 278
Kirkpatrick, Mark 218
kiwis 74, 261
Klinefelter syndrome 129
Kokko, Hanna 39
Komdeur, Jan 263
Komodo dragons 86, 176
Koskella, Britt 61

Lamarck, Jean-Baptiste 76
Lamarckism 76
Lande, Russell 218
Langlois, Judy 223, 227
langur monkeys 242

Index

Latin American populations 276–7, *277*, 283
'law of constant extinction' 41
lek 270–71, 301
Levit, Solomon 20
Liars of Nature and the Nature of Liars, The (Sun) 243
life cycles 78–9, *79, 80*, 301
life on Earth, 68–70
Lincoln, Abraham 229, *230*
lions 179, 242
Lively, Curt 44–50, 57–60, 64–6, 222
lizards 176
   brown anole lizard 262
   side-blotched lizards 98
   whiptail lizards 33–4
Lummaa, Virpi 148–9
Lysenko, Trofim 20

Maasai, Kenya 275
macaques 155, 198
Madame Bovary (Flaubert) 231–3, 253
Madame Bovary's Ovaries (Barash & Barash) 233
Madison, James 229
major histocompatibility complex (MHC) system 211
males and females 13, 70–71, 73–4, 81–3, 86–8, 90, 96–9, 162–6
   'battle of the sexes' 70–73, 176, 284
   secondary sexual characteristics 90–91
   sex chromosomes 104–8, *105, 106*, 120–25, 129–35, 248–50, 302
   sex identity 131–5
   sex ratios 258–80, *259*
   sexual conflict 231–55
   see also gender; mate competition; sex determination
Malheur National Forest, Oregon 81

mallards 273
mammals 11n, 59, 95, 105–7, *105*, 121, 124, 129–30, 149, 235n, 241–2, 248, 263
   homosexuality and 155, 158
   mate choice 211–12, 214, 248–9
   mate competition 179, 188, 191
   placental mammals 250
   reproduction 251, 260
   sex ratio 263, 264–7
manakins 270
mandrills 214–15, *215*, 217
marine life, evolution of 96
marriage 226, 232, 252
marsupials 249–50
Martin, Isaac 172–3
master genes 120–23, 302
master-switch genes 120–23
mate choice 179, 194, 202–230, 273–4, 291, 302
   beauty 202–5, 207–11, 226–30, 273–4
   birds 209–12
   bright colours and 209, 210, 214–15, 217
   female choice 202–12, 222
   fertilization and 211–12
   genetic linkage and 218–19
   handicap hypothesis 219–20
   human mate choice 223–30, 273–4
   indirect benefits 208–9
   male choice 212–14
   mammals 211, 212
   mating calls 215–16, 217
   MHC genes and 211
   reduction of variation 226
Mate Choice (Rosenthal) 226
mate competition 174–99, 302
   body size 179
   conflicts 180–82
   cooperative bargaining 182

evolved weapons 177, 178–80, 188, 188
female competition 189–91
humans and 194–8
infanticide 183, 191
male competition 175–89\
mate guarding 192–4
social contracts 182–4
sperm production 184–9, 185
mating types 2, 83–6, 302
Max Planck Institute 187
mayflies 260
Maynard Smith, John 13
McCoy, William 172–3
meadow voles 187
meerkats 191
meiosis 13, 14, 22n, 34, 80, 302
Mesoamerica 278
Mesopotamia 278
mice 76, 90–91, 93, 151, 191, 247, 249–51, 265
Microphallus trematodes, 47
Miller, Geoffrey 225
Mills, John 172–3
mitochondria 23–4
mitosis 12
moles 3, 150, 154
Mongolian gerbils 147–8,
Mongols 276
monkeys 127, 191
  langur monkeys 242
  uakari monkeys 215, 217
monogamy 234–6, 234, 252, 279
  genetic monogamy 236
  social monogamy 236, 279
Montgomerie, Robert 203
moose 176, 269
Morgan, Thomas Hunt 17
Morin, Peter 49
Mormon crickets, 190

mosquitoes 260
Mukogodo people, Kenya 275
Muller, Hermann Joseph 16–22, 49, 205
Muller's ratchet 16–17, 21–32, 34–6, 286–7, 302
Muscovy ducks 244–5
mushrooms *see* fungi
mutation 21–3, 35, 39, 43, 286
  'mutation load' 21
  'mutational meltdown' 22
mycelia 79, 81

naked mole-rats 110, 112, 126–7
narwhals 181, 182, 203
natural selection 29, 89, 202
Nature journal 64
nematodes 32, 38, 47, 52–3, 59
  Strongyloides ratti 59
Neotrogla cave insects 142–5, 150, 190
neuroligins 153
'New Evolutionary Law' (Ridley) 40
New Mexico, USA 33–4
Nietzsche, Friedrich 69
'noble savages' 182
northern cardinals 209
Norway 149
Nosema whitei parasite 47

octopuses 155
oestrogen 146, 148
Olympic Games: 133 (1996); 134 (2024)
opossums 263
orangutans 243
ostriches 74, 139
Otto, Sarah (Sally) 28–9, 32, 64–6, 222
Out of the Night (Muller) 20
Ovid 128
ovotestes 3

Index

painted snipes 264
Panama 207
parakeets 211
parasites 42–8, 44, 53–4, 59–61
parrots 210
parthenogenesis 11, 33–4, 86, 302
parthenogenic triploid (3n) fish 49
pathogens 42, 44, 47–8
peafowl 202–5, 204, 208, 218–22, 262, 268–9
penguins 139, 155
Peromyscus deer mice 186
pesticide pollution 146–7
Peters' dwarf frogs 217–18
Petrie, Marion 203–4, 208, 219, 222, 262
phalaropes 264
Phelps, Steve 222
Philadelphia (film) 166
Philoponella prominens spider 15
pigeons 10
 passenger pigeons, 56
 rock pigeons 266
Pinker, Steven 278
pipefish 141, 213
Pitcairn Island 172–5, 196, 198, 234
plants 11, 23n, 31n, 32, 42, 78, 81n, 96, 121n, 265
plasmids 25
plastic surgery 201
Plato 128
platyfish (*Xiphophorus maculatus*) 258–9, 259
platypus 103–4, 106, 106, 124
Pliny 128
Poeciliopsis lucida fish 49
Poeciliopsis monachal fish 49
polygamy 269
polygyny 233, 236–8, 252, 260, 262, 274–5, 278

Potamopyrgus antipodarum mud snail 46–7, 57–61
praying mantises 15
predator-prey interaction 61–2, 62
'pretty privilege' 227–8
primates 216–17, 242
princess pygmy goby 235
prkar1a gene 186
progesterone 267
pronghorns 273
protists 83
proto-sexes 2, 84, 85
protozoans 1–2, 84
 Tetrahymena thermophila 1–2, 1, 84
Prum, Richard 131

Quintal, Matthew 172–3

r-strategists 54–5, 301
rape 254
rats 59
rattlesnakes 180, 180, 182
ravens 210
red deer 181, 182, 234, 247, 251, 264
Red Queen, The (Ridley) 285
Red Queen hypothesis 41–67, 96, 284, 287, 302
red-eared slider turtle 114
Reflections of Eden (Galdikas) 243
Reichard, Ulrich 187
'reproductive flip-flopping' 29–34, 30, 38–9, 53–6, 80–81, 287
reptiles 113–15
 see also names of individual reptiles
Rice Institute, Houston 18
Richardson, Sarah 152
Rickettsia bacteria 24
Ridley, Matt 40, 285
Roberts, Julia 243
Robertson, Raleigh 268

rock pigeons 266
rodents 242, 246–7
Roman Empire 128, 163
Röntgen, Wilhelm 18
roosters 176
Rosenthal, Gil 209, 222, 226
rotifers 25–6, 27, 29, 43, 53, 62
   bdelloid rotifers 25–6, 27, 29, 43, 53
Rousseau, Jean-Jacques 182, 183–4, 196
Rubio, Marco 229
ruff sandpipers 98, 98
Rwanda 275
Ryan, Mike 205–8, 215–17, 221–3, 227

Saccharomyces yeast 85
Sagan, Carl 206
salmon 15, 73–4, 98
Samburu, Kenya 275
San Diego Zoo, California 9–10
scarab beetles 188, 188
Schlupp, Ingo 222
Schmitt, David 224–5
scissortail sergeants 213
seahorses 139–41, 161, 161, 176, 213, 235, 238
seals 155
Selfish Gene, The (Dawkins) 88–9
Semenya, Caster 131–2, 134
sensory exploitation hypothesis 215–17, 302
Serratia marcescens 47
sex abstinence 38–9
Sex and Evolution (Williams) 45
sex determination 4, 104–35, 289, 302
   bending gender 4, 139, 143–4, 160
   environmental 113–17
   genetic 104–13, 120–25
   sex change 118–20, 141, 158–60
   sex identity 131–5, 289
   social 117–19

sex identity 131–5, 289
Sex Is a Spectrum (Fuentes) 165
sex ratio 258–80, 259, 303
   genetics and 265–6, 272, 276–8
   hormones and 266–8
   humans and 273–80
   nutrition and 267–8
   violence and 278–79
sex reversal 118–20, 141, 289, 303
sex-selective abortion 279
sexual conflict 231–55, 234, 291, 303
   coercion 243–4
   genetics and 245–51, 247
   hormones and 241
   humans and 251–4
   infanticide and 242, 254–5
   monogamy and 234–6, 234, 241–2, 252
   parental care and 237–40
   paternity 238–9, 242
   placenta and 250
   polygamy and 236–8
   sexual deceit 243
sexual dimorphism 194, 203, 214, 251, 291
sexual orientation 151–8, 166–7
   bisexuality 157–8
   evolution and 156–7
   genes and 152
   homosexuality 151–5, 166–7
   hormones and 154–5
   kin selection and 156–7
   non-social environmental factors 152–4
   prenatal environment 154–5
   same-sex preference 156–8
sexual relationships 15–16
sexual reproduction 11–15, 29–31, 62–3, 286–94
   condition-dependent sex 32–3

Index

'cost of meiosis' 13, 108–10, 112, 126
'cost of sex/cost of males' 13–14, 299–300
   debut of 70–72
   diversity and 95–7
   genetic recombination 17, 32, 35, 65, 88–9, 92–4, 122, 174, 280–83, 287–8
   regular sex 39–44, 55–6
   social cost of 109–10, 126, 288, 303
   two-sex system, 82–4, 86–8
   'twofold cost' 13–14, 26, 55, 126, 286
   see also males and females
sexual selection see mate choice
'sexy-son' hypothesis 268–73, 276–7, 303
Seychelles warblers 193, 263–4
sharks, bonnethead 86
Shaw, George 103
sheep 154–5, 246
shrews 242
shrimp 64–5
side-blotched lizards 98
single-celled organisms 83
slime moulds 83, 84
Smith, John Maynard 26
smooth guardian frogs, 239
snails 46–7, 57–61, 86
   cloning 57–8
   mud snails 46–7, 57–61
snakes 180
   garter snakes 186, 192
   rattlesnakes 180, *180*, 182
snow skinks 115–16, *116*
Soay, Scotland 246
Social Contract, The (Rousseau) 182
social cost of sex 109–10, 126, 288, 303
social sex determination 117–18
Solter, Davor 250

soma-germ division 77–8, *77*, 91
Song, Siliang 157–8
South Korea 201, 279
Soviet Union 19–20
   Great Purge (1936–8) 19–20
sperm 73–4, 83, 86–7, 90, 93, 97, 176–8, 184–9, *185*, 205
   ejaculates 213
   sperm competition 185–9, 192, 194, 212, 245, 253, 303
spiders 15, 192, *193*
   Philoponella prominens 15
spotless starlings 240
spotted hyenas 143–5, *145*, 213
SRY gene 123–4, 133–5, 149
Stalin, Joseph 19–20
stalk-eye flies 219
starlings, 236
   spotless starlings 240
stick insects (genus *Timema*) *30*, 31
sticklebacks 208–9, 219, 246
Strongyloides ratti nematode 59
Surani, Azim 250
Surinam toads 238–9, *239*
swallows 192
swamp eels (*Monopterus albus*) 118–20, *118*, 141, 159
Sylvestre, Florent 246

Takahashi, Mariko 203
tardigrades (water bears) 53–4
technological limitation 292
teleonomy 89–90, 303
terns 155
testosterone 133–5, 146, 148–51, 154–5, 266–8, 224
Tetrahymena hyperangularis ciliate 84
Tetrahymena thermophila protozoa 1–2, *1*, 84
Texas, USA 18–19

theoretical biology 65
Thogerson, Collette 263
Through the Looking-Glass (Carroll) 41
Tibetan macaques 270
tits
    Eurasian penduline tits 239–40, *241*
    great tits 194
Tocqueville, Alexis de 166–7
topminnows 49–50
transduction 24, 28, 303
transformation 24, 28, 303
trematodes 46–7
triplefin blenny 190
triploids 49, 58, 303
trisomy 129
Trivers–Willard hypothesis 262–3,
    264n, 269, 273, 274–7, 303
Trivers, Robert 258, 261
túngara frogs 215, 221–2
Turner syndrome 129
turtles 113–14
    red-eared slider turtle 114
'twofold cost of sex' 13–14, 26, 55,
    126, 286

uakari monkeys *215*, 217
UK Biobank 157
unisexual organisms 83–5, *83*
    mating types 84–5
United States
    homosexuality and 166–7
    LGBTQ rights 292–3
    politics 88
    presidential elections 229
    sex ratios 275
University of Texas, Austin 18, 20,
    205–6, 222–3

van Valen, Leigh 40–42, 61, 63–4,
    66–7, 284

Vavilov, Nikolai 19
Vedic Valley 278
virgin birth 10–11, 86, 176
voles 187, 241
Vrijenhoek, Bob 49–50, 58

walruses 188
Wang Zhaojun 274
wasps 107, *108*, 110
water fleas 32–3, 38, 52–3, 86
Weatherhead, Patrick 268
Weber, Ernst Heinrich 220
Weber–Fechner law 220–22, *220*, 303
Weismann, August 75–8, *77*, 91
whales 155
whiptail lizards 33–4
Wilczynski, Walt 223
Will & Grace (TV series) 166
Willard, Dan 258
Williams, George 13, 45
Williams, John 172–3
Wittgenstein, Ludwig 290
wolves 52
wrasses 119–20
wrens 194

X-rays 16, 18, 21
Xishi 274

Yang Yuhuan 274
Yanomamö society, South America 196
yeasts 38, 79–81, 83, 85
    Saccharomyces 85
Yorzinski, Jessica 204
Young, Edward 172–3

Zahavi, Amotz 219, 222
Zerjal, Tatiana 276
Zhang, Jianzhi 157–8
zygotes 12, 72, 93